T0313900

Advanced Multifunctional Lightweight Aerostructures

Wiley-ASME Press Series

Advanced Multifunctional Lightweight Aerostructures: Design, Development, and Implementation
Kamran Behdinan and Rasool Moradi-Dastjerdi.

Vibration Assisted Machinery: Theory, Modelling, and Applications
Li-Rong Zheng, Wanqun Chen, and Dehong Huo

Two-Phase Heat Transfer
Mirza Mohammed Shah

Computer Vision for Structural Dynamics and Health Monitoring
Dongming Feng, Maria Q Feng

Theory of Solid-Propellant Nonsteady Combustion
Vasily B. Novozhilov, Boris V. Novozhilov

Introduction to Plastics Engineering
Vijay K. Stokes

Fundamentals of Heat Engines: Reciprocating and Gas Turbine Internal Combustion Engines
Jamil Ghojel

Offshore Compliant Platforms: Analysis, Design, and Experimental Studies
Srinivasan Chandrasekaran, R. Nagavinothini

Computer Aided Design and Manufacturing
Zhuming Bi, Xiaoqin Wang

Pumps and Compressors
Marc Borremans

Corrosion and Materials in Hydrocarbon Production: A Compendium of Operational and
Engineering Aspects
Bijan Kermani and Don Harrop

Design and Analysis of Centrifugal Compressors
Rene Van den Braembussche

Case Studies in Fluid Mechanics with Sensitivities to Governing Variables
M. Kemal Atesmen

The Monte Carlo Ray-Trace Method in Radiation Heat Transfer and Applied Optics
J. Robert Mahan

Dynamics of Particles and Rigid Bodies: A Self-Learning Approach
Mohammed F. Daqaq
Primer on Engineering Standards, Expanded Textbook Edition
Maan H. Jawad and Owen R. Greulich

Engineering Optimization: Applications, Methods and Analysis
R. Russell Rhinehart

Compact Heat Exchangers: Analysis, Design and Optimization using FEM and CFD Approach
C. Ranganayakulu and Kankanhalli N. Seetharamu

Robust Adaptive Control for Fractional-Order Systems with Disturbance and Saturation
Mou Chen, Shuyi Shao, and Peng Shi

Robot Manipulator Redundancy Resolution
Yunong Zhang and Long Jin

Stress in ASME Pressure Vessels, Boilers, and Nuclear Components
Maan H. Jawad

Combined Cooling, Heating, and Power Systems: Modeling, Optimization, and Operation
Yang Shi, Mingxi Liu, and Fang Fang

Applications of Mathematical Heat Transfer and Fluid Flow Models in Engineering and
Medicine
Abram S. Dorfman

Bioprocessing Piping and Equipment Design: A Companion Guide for the ASME BPE Standard
William M. (Bill) Huitt

Nonlinear Regression Modeling for Engineering Applications: Modeling, Model Validation,
and Enabling Design of Experiments
R. Russell Rhinehart

Geothermal Heat Pump and Heat Engine Systems: Theory and Practice
Andrew D. Chiasson

Fundamentals of Mechanical Vibrations
Liang-Wu Cai

Introduction to Dynamics and Control in Mechanical Engineering Systems
Cho W.S. To

Advanced Multifunctional Lightweight Aerostructures

Design, Development, and Implementation

Kamran Behdinan and Rasool Moradi-Dastjerdi
University of Toronto, Toronto, Canada

This Work is a co-publication between John Wiley & Sons Ltd and ASME Press.

WILEY

Registered Office

John Wiley & Sons, Inc., 111 River Street, Hoboken, NJ 07030, USA

Editorial Office

111 River Street, Hoboken, NJ 07030, USA

For details of our global editorial offices, customer services, and more information about Wiley products visit us at www.wiley.com.

Wiley also publishes its books in a variety of electronic formats and by print-on-demand. Some content that appears in standard print versions of this book may not be available in other formats.

Library of Congress Cataloging-in-Publication Data

Names: Behdinan, Kamran, 1961- editor. | Moradi-Dastjerdi, Rasool, 1984-
    editor. | John Wiley & Sons, Inc., publisher.
Title: Advanced multifunctional lightweight aerostructures : design,
    development, and implementation / Kamran Behdinan and Rasool
    Moradi-Dastjerdi, University of Toronto.
Other titles: Wiley-ASME Press series.
Description: First edition. | Hoboken, NJ : John Wiley & Sons, Inc., 2021.
    | Series: Wiley—ASME Press Series | Includes bibliographical references
    and index.
Identifiers: LCCN 2020033775 (print) | LCCN 2020033776 (ebook) | ISBN
    9781119756712 (cloth) | ISBN 9781119756729 (adobe pdf) | ISBN
    9781119756736 (epub)
Subjects: LCSH: Airplanes—Design and construction. | Lightweight
    construction. | Aerospace engineering.
Classification: LCC TL671.2 .A3195 2021 (print) | LCC TL671.2 (ebook) |
    DDC 629.134/1–dc23
LC record available at https://lccn.loc.gov/2020033775
LC ebook record available at https://lccn.loc.gov/2020033776

Cover Design: Wiley
Cover Image: © guvendemir/Shutterstock

Set in 9.5/12.5pt STIXTwoText by SPi Global, Chennai, India
Printed and bound by CPI Group (UK) Ltd, Croydon, CR0 4YY

C9781119756712_160421

Professor Kamran Behdinan affectionately dedicates this book to his wife, Nasrin, without her love, patience and sacrifices, this and much else would not be possible, and to his dear daughters, Dr. Tina Behdinan and Dr. Asha Behdinan, for their love and unwavering support.
Dr. Rasool Moradi-Dastjerdi warmly dedicates this book to his wife, Arezou, and to his parents for their encouragement, love, and unlimited support.

Contents

Preface

In the aerospace industry, innovative designs, which can simultaneously address concerns about safety and fuel efficiency, have created demands for novel materials and structures. These demands have persuaded researchers to propose and investigate advanced structures made of multifunctional lightweight materials. Some notable mentions include porous, composite, nanocomposite, ultra-high temperature ceramic, piezoelectric, and functionally graded materials.

In the design of aerostructures, strength to weight ratio is the key point where the use of lightweight materials results in a considerable reduction in structural weight. However, the decrease of structural strength usually comes with an ordinary reduction in the weight of the utilized materials such as embedding porosity, or use of lighter materials. This reduction in the structural strength can be compensated using multifunctional materials or by improving the design of the structures. These facts were the key drivers in the development of new lightweight technologies where traditional composite materials as lightweight materials have seen greater integration into aerostructure applications over the past two decades. Furthermore, the introduction of nanotechnology into the design of composite materials presents another leap in the increasing effort to reduce weight and tailor the material properties to suit specific aerostructure applications. In this new generation of composites, nanoscale fillers highly affect the overall properties of the resulting nanocomposite materials.

Another class of multifunctional materials which can have specific applications in the aerospace industry are piezoelectric materials. Employing piezoelectric components with the ability to convert electrical charge to mechanical load or vice versa provides self-controlling property with fast response for the whole structure. This converting ability also provides the benefit of harvesting energy, strain measurements, and damage detection in the structures.

Moreover, aerostructures are mainly subjected to either mechanical or thermal loads where ceramic materials can be prospective candidates. Among them, ultra-high temperature ceramics are an advanced class of material that experience superior structural and thermal stability, reaching temperature over 3000 °C without a noticeable sacrifice in strength.

In aerostructures, the use of architected structures along with multifunctional lightweight materials has opened up possibilities for designs previously unimaginable. The analysis of such advanced materials and structures necessitates the application of novel methods which are precise, reliable, and computationally efficient. The necessity of utilizing advanced methods complement complex geometrical shapes, different applied loads, a multi-physics environment, and a wide range of scales from nano up to macro scales.

To cover recent developments about the aforementioned concerns and their most exciting aspects, this book is divided into two parts with 10 chapters overall where the state of the art in the respective fields are comprehensively discussed.

The first part deals with multi-disciplinary modeling and characterization of some advanced materials and structures by developing new methods. This part is composed of four chapters. Specifically, layer arrangement impact on deflections of a proposed five-layer smart sandwich plate subjected to electromechanical loads is investigated in Chapter 1. The layers of the sandwich plates are assumed to be made of three different advanced multifunctional materials including porous, graphene-reinforced nanocomposite, and piezoelectric materials. In Chapter 2, heat transfer analysis of sandwich cylinders consisting of a polymeric core and functionally graded graphene-reinforced faces is studied. Chapter 3 presents the application of a new multiscale approach in the modeling and design of lightweight materials and structures that demonstrate complex phenomena that span multiple spatial and temporal scales. In Chapter 4, chemical kinetics and the multiscale characterization of crystallinity in ultra-high temperature ceramics, polymorphic structures and their composites are described.

In the second part of this book, behaviors of some critical parts of aircraft are discussed as practical benchmark problems. Chapter 5 presents an optimization study on the design of a novel shimmy damper mechanism for aircraft nose landing gears as a practical case in aerostructures. In Chapter 6, a widely used component of helicopters, jet engines and aircraft, a flexible rotor supported by viscoelastic bearings, is modeled to study its nonlinear dynamic behavior. Chapter 7 proposes an innovative analytical–numerical method to efficiently predict the real-time temperature of belt drive systems. Chapter 8 develops an efficient and reliable physics-based approach to predict noise at the far-field of a nose landing gear of an aircraft. Another important part of aircraft is the aeroengine. In the aeroengine, the vibrations and noises can be transferred to the aircraft fuselage and this transmissibility significantly affects the aircraft crew and passenger comfort and safety. In this regard, Chapter 9 develops and implements a reliable analytical transmissibility method to analyze and investigate the vibration energy propagation in an aeroengine structure. This method results in design guidelines which can significantly reduce the development costs as well as the ability of addressing noise and vibration problems in the structure. Chapter 10 also utilizes the developed method in Chapter 9 to perform structural health monitoring to detect and classify the importance of defects and damage in the considered aeroengine using the obtained frequency response functions. Accordingly, this chapter proposes design guidelines which significantly improve the reliability and operational lifetime of the aeroengine at the lowest possible cost.

This book delivers extensive updated investigations and information to address the latest demands for the effective and efficient design and precise characterization of advanced multifunctional lightweight aerostructures. The authors believe that it is a comprehensive and useful reference for graduate students who want to increase their knowledge. This book provides innovative and practical solutions for active engineers, especially in the aerospace industry, who are looking for alternative materials, structures or methodologies to solve their current problems.

All contributions to this book are the result of years of research and development conducted by the research team under the direct supervision of the principal investigator, Professor Kamran Behdinan, in the Advanced Research Laboratory for Multifunctional Lightweight Structures (ARL-MLS) at the University of Toronto. We would like to acknowledge the funding received from the Canadian Foundation for Innovation as well as the Natural Science Engineering Research Council of Canada in support of the ARL-MLS facilities and training of highly qualified personnel. Furthermore, we wish to take this opportunity to sincerely express our appreciation to the ARL-MLS graduate students and postdoctoral fellows for their outstanding research in addressing problems of utmost significance to the aerospace research community/industry in the field of advanced multifunctional lightweight aerostructures. Their informative contributions have allowed our ideas and dreams to become a reality in this book. We are also grateful to Wiley and its team of editors for helping us to finalize this book.

Toronto, June 2020

Kamran Behdinan
Rasool Moradi-Dastjerdi

Biographies

Professor Kamran Behdinan

Professor Behdinan earned his PhD in Mechanical Engineering from the University of Victoria in British Columbia in 1996 and has considerable experience in both academic and industrial settings. He was appointed to the academic staff of Ryerson University in 1998, tenured and promoted to the level of associate professor in 2002 and subsequently to the level of professor in 2007 and served as the director of the aerospace engineering program (2002–2003), and the founding Chair of the newly established Department of Aerospace Engineering (July 2003–July 2011). Professor Behdinan was a founding member and the Executive Director of the Ryerson Institute for Aerospace Design and Innovation (2003–2011). He was also a founding member and the coordinator of the Canadian–European Graduate Student Exchange Program in Aerospace Engineering at Ryerson University. He held the NSERC Design Chair in "Engineering Design and Innovation," 2010–2012, sponsored by Bombardier Aerospace and Pratt and Whitney Canada. He joined the Department of Mechanical and Industrial Engineering, University of Toronto, as a professor in September 2011. He is the NSERC Design Chair in "Multidisciplinary Design and Innovation – UT IMDI," sponsored by NSERC, University of Toronto, and 13 companies including Bombardier Aerospace, Pratt and Whitney Canada, United Technology Aerospace Systems, Magna International, Honeywell, SPP Canada Aircraft, Ford, and DRDC Toronto. He is the founding director of the "University of Toronto Institute for Multidisciplinary Design and Innovation," an industry-centered, project-based learning institute in partnership with major aerospace and automotive companies.

Professor Behdinan is a past President of the Canadian Society of Mechanical Engineering (CSME), and served as a member of the International Union of Theoretical and Applied Mechanics (IUTAM) – General Assembly and the IUTAM Canadian National Committee, and a member of technical and scholarship committees of the High-Performance Computing Virtual Laboratory (HPCVL). He is the founding director and principal investigator of the University of Toronto, Department of Mechanical and Industrial Engineering "Advanced Research Laboratory for Multifunctional Lightweight Structures," funded by the Canadian Foundation for Innovation as well as Ontario Research Fund. His research interests include Design and Development of Light-weight Structures and Systems for biomedical, aerospace, automotive, and nuclear applications, Multidisciplinary Design Optimization of Aerospace and Automotive Systems, Multi-scale Simulation of

Nano-structured Materials and Composites. He has supervised 32 PhDs, 120 Masters, and 40 Post-Doctoral Fellows and Scholars. He has also published more than 370 peer-reviewed journal and conference papers, and 9 book chapters. He has been the recipient of many prestigious awards and recognitions such as the Research Fellow of Pratt and Whitney Canada and Fellows of the CSME, ASME, the Canadian Academy of Engineering, EIC, AAAS, as well as Associate Fellow of AIAA.

Dr. Rasool Moradi-Dastjerdi

Dr. Moradi-Dastjerdi is currently a postdoctoral fellow in the Advanced Research Laboratory for Multifunctional Lightweight Structures at the University of Toronto. He obtained his PhD degree in Mechanics at Shahid Rajaee Teacher Training University, Tehran, Iran in 2016.

His research focuses on the coupled thermo-electro-mechanical analysis of smart multifunctional lightweight structures made of advanced materials such as piezoelectric materials, nanocomposites, functionally graded materials, and foams. He mainly utilizes advanced numerical methods including mesh-free and finite element methods. He has been an advisor for two PhD and six MSc theses. He has also contributed to more than 50 peer-reviewed journal papers. He was the recipient of the best researcher award from the Young Researcher and Elite Club of Isfahan Azad University in 2011, 2012, 2013, and 2016.

Part I

Multi-Disciplinary Modeling and Characterization

1

Layer Arrangement Impact on the Electromechanical Performance of a Five-Layer Multifunctional Smart Sandwich Plate

Rasool Moradi-Dastjerdi and Kamran Behdinan

Advanced Research Laboratory for Multifunctional Lightweight Structures (ARL-MLS), Department of Mechanical & Industrial Engineering, University of Toronto, Toronto, Canada

1.1 Introduction

The reversible effect of piezoelectricity is the ability to generate electrical charge as a result of subjecting to mechanical loads. This active effect is observed in some specific materials called piezoelectric materials. Such active materials are usually employed as attachments or layers in passive structures to provide a self-controlling property with fast response in the resulting smart structures [1]. The application of such active materials mainly relies on their passive structures. In the design of aerostructures, weight and strength are two key points which can be addressed using sandwich structures as they generally contain a thick lightweight core for stabilizing the structures and two thin stiff faces to provide structural strength [2–4]. In this regard, attaching thin layers made of polymer base nanocomposites onto the faces of polymeric porous cores results in multifunctional sandwich structures [5]. Moreover, the use of such passive structures as the host of piezoceramic attachments reduces failure risks due to the brittle structure of piezoceramics. Depending on the application, a wide range of nanofillers with astonishing thermomechanical properties have been proposed and utilized in nanocomposites. Among them, the extraordinary nanofillers of graphene and carbon nanotubes (CNTs) have aroused the interest of researchers in both academia and industry [6–8]. Although there are different parameters in the electromechanical design of five-layer multifunctional smart sandwich plates (5LMSSPs), protecting brittle layers of piezoceramic is also an important issue. Changing the location of piezoceramic layers from faces to middle layers (i.e. between porous core and nanocomposite faces) provides protecting layers. This change in layer arrangements can affect both the mechanical and electrical response of such structures.

In recent literature, different passive structures have been considered as host for piezoelectric sensors/actuators to introduce smart structures with potential applications in energy harvesters [9], noise and vibration reduction [10], fluid delivery [11, 12] and structural damage monitoring [13] where piezoelectricity plays an essential role. In these structures, piezoelectric components come as attached pieces, separate layers or (nano)fillers. Askeri et al. [14] proposed attaching two lead zirconate titanate (PZT) layers on the faces of a transversely isotropic nonpiezoelectric plate to introduce a smart plate. Functionally graded (FG) metal/ceramic materials, as advanced materials, have been considered as the passive part of smart structures activated by piezoelectric materials. In this regard, passive plates and shells made of such FG materials and PZT-activated layer(s) were considered under thermo-electro-mechanical loads to study their nonlinear dynamic responses in [15–17]. The deflections of FG titanium/aluminum oxide plates integrated between PZT faces under static and dynamic electromechanical loads were presented in [18]. Khoa et al. [19] covered the outer layer of imperfect FG metal/ceramic cylindrical shells with a PZT layer and studied its buckling resistance. In another setting, laminated composites have been also employed as the passive part of piezoelectric activated smart structures. Talebitooti et al. [20] considered such plates covered with PZT sensor and actuator layers to optimally control the vibrations of the obtained smart plates using a feedback algorithm. By developing an isogeometric finite element method (FEM), Phung-Van et al. [21] aimed to outline the static deflections and vibrations of the same composite plates actuated by PZT layers. To reduce structural weight, nanocomposite materials have also been used as the host of piezoelectric actuated smart structures. In addition, there are different types of nanofillers that can be utilized in specific applications. The use of composite plates enhanced with wavy CNTs and carbon fibers as the multifunctional host of two piezoelectric patches was proposed by Kundalwal et al. [10]. They utilized piezoelectric patches made of piezoceramic fibers embedded in a polymer to provide smart damping property for host plates. Mohammadimehr et al. [22] employed nanofillers of CNTs and piezoelectric nanotubes of boron nitride in passive and active polymers and considered double sandwich plates. They presented the vibration behaviors of such smart structures subjected to magnetic and electric fields. Moradi-Dastjerdi et al. [23] proposed the use of nanocomposite plates enhanced with nanoclays in aggregated and intercalated forms as the passive layers of a smart plate with two PZT faces. Arani et al. [24] utilized two piezoelectric faces made of polyvinylidene fluoride to control the frequencies of CNT/polymer microplates under magnetic field and located on an elastic foundation. Malekzadeh et al. [25] considered a graphene/polymer multi-layered circular plate with a randomly located hole activated with two PZT faces and outlined its vibration behavior. For further reduction of structural weight in the passive layer, porous material can be utilized. Jabbari et al. [26, 27] suggested the use of circular plates with FG dispersions of embedded porosities for passive layers activated by attaching two PZT faces and investigated the stability resistances of the obtained smart circular plates. Askari et al. [28] considered rectangular plates with FG patterns of porosity dispersion between two PZT faces to determine porosity impact on the natural frequencies of the resulted active plates. Barati and Zenkour [29] studied the vibrations of active porous plates made of an FG mixture of two different piezoceramics. Mohammadi et al. [30] considered aluminum cylindrical pressure vessels with three patterns of porosity dispersion integrated between two inner and outer PZT faces as sensor and actuator. They presented the electromechanical responses of such smart pressure vessels located in elastic media. In a more advanced setting of smart structures, the combination of nanocomposite and porous materials have been utilized as multifunctional passive structures. Nguyen et al. [31, 32] proposed three-layer smart sandwich plates with a metal porous layer enhanced with graphene platelets (GPLs) as a passive core integrated between two active PZT faces. They considered FG patterns for the dispersions of GPLs and porosities in the passive layer and obtianed two different sets of results including vibrational response and its active control using PZT layers. However, GPLs in the core layer of these three-layer smart plates interfere with the electrical charge and potential field obtained in PZT faces. In addition, according to the concept of sandwich panels, the use of separated layers of nanocomposite and porous materials leads to five-layer multifunctional smart panels with higher structural stiffness to weight ratio. In these regards, Setoodeh et al. [33] proposed such five-layer smart curved shells including two PZT faces, two CNT-enhanced nanocomposite middle layers and one porous core with FG patterns for the dispersions of nanofillers and porosities. Another set of five-layer smart plates including PZT faces, graphene/polymer middle layers and porous core were also proposed and thermo-electro-mechanical behaviors of such 5LMSSPs are presented in [34–37].

Five-layer multifunctional smart sandwich plates with layers made of porous, GPL/polymer and PZT have been proposed in the literature. However, considering piezoceramic layers as the faces of such plates is a challenging point because of their brittle nature. Therefore, in this work, a comparison study has been conducted to explore the impact of changing the location of piezoceramic layers from faces to middle layers on the electromechanical performances of 5LMSSPs. In this regard, a mesh-free solution has been developed based on Reddy's third-order shear deformation theory (TSDT). Moreover, Halpin–Tsai equations with the ability to capture the shape of nanofillers are employed to define the mechanical properties of GPL/polymer nanocomposite layers. In addition to the impact of layer arrangement, the effects of GPL volume and dispersion, porosity volume and the thickness of each layer are investigated in this chapter.

1.2 Modeling of 5LMSSP

The considered multifunctional smart sandwich plates have five layers including one porous, two piezoelectric and two GPL-reinforced nanocomposite layers to provide a wide range of industrial applications. There is no doubt in the use of a porous layer as the core because embedding porosity in the core results in a remarkable structural weight reduction without a significant loss of structural stiffness. However, the locations of nanocomposite and piezoelectric layers could be changed based on the operating conditions of 5LMSSPs. As shown in Figure 1.1, two different layer arrangements have been examined: case I, considering piezoceramics as faces; and case II, considering nanocomposite layers as the faces of 5LMSSP to protect the piezoelectric from environmental and loading issues. In both cases, 5LMSSPs are assumed under a uniform mechanical pressure f_0 on their top faces. In addition, the piezoelectric layers are subjected to an electrical input such that their outer faces are connected to a uniform voltage V_0 and their inner faces are grounded to provide a voltage difference of V_0 through the thickness of each piezoelectric layer. In this chapter, square 5LMSSPs with side length of a and thickness of h are considered. The thicknesses of porous, piezoelectric and nanocomposite layers are represented by h_c, h_p, and h_n, respectively.

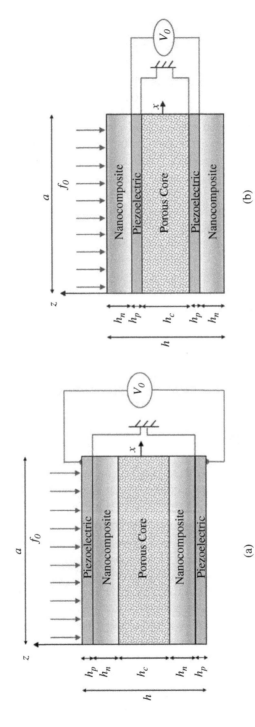

Figure 1.1 Two different layer arrangements of the considered five-layer sandwich plates with one porous, two piezoelectric and two GPL-reinforced nanocomposite layers: (a) 5LMSSP case I; and (b) 5LMSSP case II.

1.2.1 Porous Layer

Embedding pores in a body affects the material properties of the host body. In this chapter, the distribution of pores is considered as a symmetric profile through the thickness of the core layer such that the outer faces of the core i.e. $z = \pm h_c/2$ has no pores, but the highest volume of pores is located at $z = 0$. Using Gaussian random field technique [38], Young's modulus E^p, density ρ^p and Poisson's ratio υ^p of such porous layer can be estimated as [31]:

$$E^p(z) = \left(1 - q_0 \cos\left(\frac{\pi z}{h_c}\right)\right) E^m \tag{1.1}$$

$$\rho^p(z) = \left(1 - q_m \cos\left(\frac{\pi z}{h_c}\right)\right) \rho^m \tag{1.2}$$

$$\upsilon^p(z) = 0.221\beta + \upsilon^m(0.342\beta^2 - 1.21\beta + 1) \tag{1.3}$$

where superscripts m and p show the corresponding properties of perfect polymeric ($q_0 = 0$) and porous ($q_0 \neq 0$) core layer, respectively. In addition, q_0 is the porosity parameter which implies the porosity volume fraction of the core layer. q_m and β are determined as [31]:

$$q_m = \frac{1.121(1 - \sqrt[2.3]{1 - q_0 \cos(\pi z/h_c)})}{\cos(\pi z/h_c)} \tag{1.4}$$

$$\beta = 1 - \rho^p/\rho^m = q_m \cos(\pi z/h_c) \tag{1.5}$$

1.2.2 Nanocomposite Layers

Graphene platelets are assumed to be dispersed with functionally graded patterns in nanocomposite layers to optimize the volume of GPLs and to improve the structural performance of 5LMSSPs. The FG patterns of nanofillers in cases I and II can be described as the functions of GPL volume fraction V_r versus z location as follows [35]:

Case I:

Upper nanocomposite layer: $V_r(z) = [(2z - h_c)/2h_n]^p \times V_r^*$ (1.6)

Lower nanocomposite layer: $V_r(z) = [-(2z + h_c)/2h_n]^p \times V_r^*$ (1.7)

Case II:

Upper nanocomposite layer: $V_r(z) = [(2z - 2h_p - h_c)/2h_n]^p \times V_r^*$ (1.8)

Lower nanocomposite layer: $V_r(z) = [-(2z + 2h_p + h_c)/2h_n]^p \times V_r^*$ (1.9)

where V_r^* is the specific GPL volume fraction and p is the exponent of GPL volume fraction which controls the profile of GPL dispersion. The dispersion profiles of GPLs through the thickness of the considered 5LMSSP are illustrated in Figure 1.2.

To determine the effective mechanical properties of nanocomposite layers, Halpin–Tsai's approach [39, 40] capable of considering the rectangular shape of GPLs is employed.

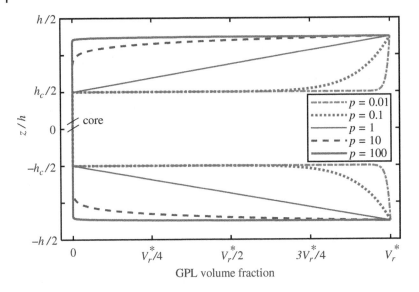

Figure 1.2 The dispersion profiles of GPLs through the thickness.

According to this approach, the effective Young's modulus of GPL/polymer nanocomposite is estimated as follows [39, 40]:

$$E(z) = \frac{3}{8}\left[\frac{1 + 2(a^g/h^g)\gamma_{11}^g V_r}{1 - \gamma_{11}^g V_r}\right]E^m + \frac{5}{8}\left[\frac{1 + 2(b^g/h^g)\gamma_{22}^g V_r}{1 - \gamma_{22}^g V_r}\right]E^m \quad (1.10)$$

where

$$\gamma_{11}^g = \frac{E^g/E^m - 1}{E^g/E^m + 2a^g/h^g}, \gamma_{22}^g = \frac{E^g/E^m - 1}{E^g/E^m + 2b^g/h^g} \quad (1.11)$$

where E^g and E^m are the Young's moduli of GPL and polymeric matrix, respectively. Moreover, a^g, b^g, and h^g are the geometrical dimensions of GPL, respectively.

1.2.3 Governing Equations

To calculate the energy function of the considered 5LMSSPs, displacement field is needed to be defined by a plate theory. It is well established that the increase of the order of plate theory makes the theory more applicable to thicker plates. Therefore, an efficient TSDT reported by Reddy [41] is utilized to determine x, y and z components of displacement field (i.e. $u, v,$ and w) in 5LMSSPs as:

$$u(x, y, z) = u_0(x, y) + z\,\theta_x(x, y) + z^3 c_1(\theta_x + w_{0,x})$$
$$v(x, y, z) = v_0(x, y) + z\,\theta_y(x, y) + z^3 c_1(\theta_y + w_{0,y})$$
$$w(x, y, z) = w_0(x, y) \quad (1.12)$$

where $c_1 = -4/3h^2$ and θ_x and θ_y are the rotations of mid-plane around the y and x axes, respectively. As Eq. (1.12) shows, the utilized plate theory has only five unknowns which is important in the reduction of computational cost. According to the utilized TSDT, in-plane ε_b and out-of-plane γ strain vectors are evaluated by [41]:

$$\varepsilon_b = \varepsilon_0 + z\varepsilon_1 + c_1 z^3 \varepsilon_3 \ , \gamma = (1 + 3c_1 z^2)\gamma_0 \tag{1.13}$$

in which

$$\varepsilon_0 = \left\{ \begin{array}{c} u_{0,x} \\ v_{0,y} \\ u_{0,y} + v_{0,x} \end{array} \right\}, \varepsilon_1 = \left\{ \begin{array}{c} \theta_{x,x} \\ \theta_{y,y} \\ \theta_{x,y} + \theta_{y,x} \end{array} \right\}, \varepsilon_3 = \left\{ \begin{array}{c} \theta_{x,x} + w_{0,xx} \\ \theta_{y,y} + w_{0,yy} \\ \theta_{x,y} + \theta_{y,x} + 2\,w_{0,xy} \end{array} \right\}, \gamma_0 = \left\{ \begin{array}{c} \theta_x + w_{0,x} \\ \theta_y + w_{0,y} \end{array} \right\} \tag{1.14}$$

Moreover, the constitutive equation which describes stress vector σ as a function of strain vector ε and elastic stiffness matrix Q in passive (porous and nanocomposite) layers of 5LMSSPs is determined as:

$$\sigma = Q\varepsilon \tag{1.15}$$

However, the constitutive equation of active piezoelectric layers is coupled with electromechanical equations which describe stress σ and electrical displacement D vectors as follows [18]:

$$\sigma = Q\varepsilon - e^T E$$
$$D = e\varepsilon + kE \tag{1.16}$$

where E is the electric field vector. In addition, e and k are the piezoelectric and dielectric constant matrices. The parameters of Eqs. (1.15) and (1.16) are defined as:

$$\sigma = \{\sigma_b \ \sigma_s\}^T, \sigma_b = \{\sigma_{xx} \ \sigma_{yy} \ \tau_{xy}\}^T, \sigma_s = \{\tau_{xz} \ \tau_{yz}\}^T \tag{1.17}$$

$$\varepsilon = \{\varepsilon_b \ \gamma\}^T, \varepsilon_b = \{\varepsilon_{xx} \ \varepsilon_{yy} \ \gamma_{xy}\}^T, \gamma = \{\gamma_{yz} \ \gamma_{xz}\}^T \tag{1.18}$$

$$Q = \begin{bmatrix} Q_b & 0 \\ 0 & Q_s \end{bmatrix}, Q_b = \begin{bmatrix} Q_{11} & Q_{12} & 0 \\ Q_{12} & Q_{22} & 0 \\ 0 & 0 & Q_{66} \end{bmatrix}, Q_s = \begin{bmatrix} Q_{55} & 0 \\ 0 & Q_{44} \end{bmatrix} \tag{1.19}$$

$$e = \begin{bmatrix} e_b & e_s \end{bmatrix}, e_b = \begin{bmatrix} 0 & 0 & 0 \\ 0 & 0 & 0 \\ e_{31} & e_{32} & 0 \end{bmatrix}, e_s = \begin{bmatrix} 0 & e_{15} \\ e_{24} & 0 \\ 0 & 0 \end{bmatrix} \tag{1.20}$$

$$k = \begin{bmatrix} k_{11} & 0 & 0 \\ 0 & k_{22} & 0 \\ 0 & 0 & k_{33} \end{bmatrix} \tag{1.21}$$

$$E = -\{0 \ 0 \ V_{,z}\}^T \tag{1.22}$$

where V is the electric potential along the thickness of piezoceramic layers.

Furthermore, the total energy Π of 5LMSSPs generated by mechanical load f_u and electrical voltage f_v is determined as:

$$\Pi = \frac{1}{2} \int_\Omega -[\varepsilon_b^T \sigma_b + \gamma^T \sigma_s - E^T D] \, d\Omega + \int_A [wf_u - Vf_v] \, dA \tag{1.23}$$

where Ω and A are the volume and top face area of 5LMSSPs, respectively.

1.3 Mesh-Free Solution

The developed mesh-free solution works based on moving least square (MLS) shape functions for the approximation of domain variables and transformation approach to impose the essential boundary conditions (i.e. mechanical supports). In mesh-free methods, nodes are dispersed in the domain and boundaries as sample points for the calculation of the desired domain variable. In the considered 5LMSSP, the desired domain variable is displacement field. In addition, a background cell with Gauss points should be considered for computing stiffness and mass matrices. By predefining effective domains for each computational (Gauss) point, local stiffness and mass matrices for each point can be calculated. Moreover, global stiffness and mass matrices can be assembled from local matrices, while the local values of common points in neighboring effective domains are accumulated in the global matrices. Figure 1.3 shows the dispersions of Gauss points and nodes, two effective domains for two Gauss points and essential boundary conditions (mechanical supports).

1.3.1 MLS Shape Function

Like finite element methods, mesh-free methods approximate a function within a specific domain using an interpolation (shape) function. In this chapter, an MLS shape function is utilized to approximate the real values of the displacement field $\mathbf{U}(\mathbf{X})$ in the domain at a point $\mathbf{X}(x, y)$ as follows [42]:

$$\mathbf{U}(\mathbf{X}) = \sum_{i=1}^{m} p_i(\mathbf{x})a_i = \mathbf{P}^{\mathrm{T}}(\mathbf{X})\mathbf{a}(\mathbf{X}) \tag{1.24}$$

where $a_i(\mathbf{X})$ is the vector of variable coefficients and $\mathbf{P}(\mathbf{X})$ is base vector with m components. In this work, quadratic base vector with $m = 6$ is utilized. This base vector is defined as:

$$\mathbf{P}(\mathbf{X}) = [1, x, y, xy, x^2, y^2]^T \tag{1.25}$$

To determine the variable coefficients $a_i(\mathbf{X})$, the following weighted least-squares error norm must be minimized at any point $\mathbf{X}$:

$$J = \sum_{i=1}^{n} W(\mathbf{X} - \mathbf{X}_i)[\mathbf{P}^{\mathrm{T}}(\mathbf{X}_i)\mathbf{a}(\mathbf{X}) - \hat{U}_i]^2 \tag{1.26}$$

where $\hat{U}_i$ is the generalized values of approximated displacement field and n is node numbers in the effective domain of point $\mathbf{X}$, and W is the cubic spline weight function in this

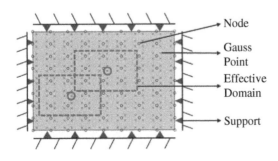

Node

Gauss
Point

Effective
Domain

Support

Figure 1.3 Node and Gauss point dispersions and effective domains in a discretized domain by a mesh-free method.

work. Minimizing J with respect to $a_i(\mathbf{X})$ introduces

$$\mathbf{a}(\mathbf{X}) = [\mathbf{M}(\mathbf{X})]^{-1} \cdot \mathbf{S}(\mathbf{X}) \cdot \hat{\mathbf{U}} \tag{1.27}$$

where

$$\mathbf{M}(\mathbf{X}) = \left[\sum_{i=1}^{n} W(\mathbf{X} - \mathbf{X}_i) \mathbf{P}(\mathbf{X}_i) \mathbf{P}^T(\mathbf{X}_i) \right]$$

$$\mathbf{S}(\mathbf{X}) = [W(\mathbf{X} - \mathbf{X}_1)\mathbf{P}(\mathbf{X}_1), W(\mathbf{X} - \mathbf{X}_2)\mathbf{P}(\mathbf{X}_2), \dots, W(\mathbf{X} - \mathbf{X}_n)\mathbf{P}(\mathbf{X}_n)] \tag{1.28}$$

Substituting Eq. (1.27) in Eq. (1.25), results in:

$$\mathbf{U} = \sum_{i=1}^{n} \psi_i \, \hat{U}_i = \psi \, \hat{\mathbf{U}} \tag{1.29}$$

where ψ_i is MLS shape function determined as:

$$\psi_i(\mathbf{X}) = \underbrace{\mathbf{P}^T(\mathbf{X})[\mathbf{M}(\mathbf{X})]^{-1} W(\mathbf{X} - \mathbf{X}_i)\mathbf{P}(\mathbf{X}_i)}_{(1\times1)} \tag{1.30}$$

1.3.2 Discretization of Domain

According to the employed plate theory defined in Eq. (1.12), the vector of displacement field $\mathbf{U}$ and its approximated vector have five independents for each node as follows:

$$\mathbf{U} = [u_{0i}, v_{0i}, w_{0i}, \theta_{xi}, \theta_{yi}]^T, \hat{\mathbf{U}} = [\hat{u}_{0i}, \hat{v}_{0i}, \hat{w}_{0i}, \hat{\theta}_{xi}, \hat{\theta}_{yi}]^T \tag{1.31}$$

Substitution of Eq. (1.29) in Eqs. (1.13) and (1.22) leads to mesh-free definitions of strain and electric field vectors as:

$$\varepsilon_b = \{ \mathbf{H}_0 + z\,\mathbf{H}_1 + c_1 z^3\,\mathbf{H}_3 \}\,\hat{\mathbf{U}}\ , \ \gamma = (1 + 3c_1 z^2)\mathbf{H}_s\,\hat{\mathbf{U}} \tag{1.32}$$

$$\mathbf{E} = -\mathbf{H}_v \hat{\mathbf{V}} \tag{1.33}$$

where $\mathbf{H}_V, \mathbf{H}_0, \mathbf{H}_1, \mathbf{H}_3$ and $\mathbf{H}_s$ are defined as:

$$\mathbf{H}_v = \begin{bmatrix} 0 & 0 & 1/h_p \end{bmatrix} \tag{1.34}$$

$$\mathbf{H}_0 = \begin{bmatrix} \psi_{i,x} & 0 & 0 & 0 & 0 \\ 0 & \psi_{i,y} & 0 & 0 & 0 \\ \psi_{i,y} & \psi_{i,x} & 0 & 0 & 0 \end{bmatrix}, \mathbf{H}_1 = \begin{bmatrix} 0 & 0 & 0 & \psi_{i,x} & 0 \\ 0 & 0 & 0 & 0 & \psi_{i,y} \\ 0 & 0 & 0 & \psi_{i,y} & \psi_{i,x} \end{bmatrix},$$

$$\mathbf{H}_3 = \begin{bmatrix} 0 & 0 & \psi_{i,xx} & \psi_{i,x} & 0 \\ 0 & 0 & \psi_{i,yy} & 0 & \psi_{i,y} \\ 0 & 0 & 2\psi_{i,xy} & \psi_{i,y} & \psi_{i,x} \end{bmatrix}, \mathbf{H}_s = \begin{bmatrix} 0 & 0 & \psi_{i,x} & \psi_i & 0 \\ 0 & 0 & \psi_{i,y} & 0 & \psi_i \end{bmatrix} \tag{1.35}$$

Introduction of the mesh-free and vector forms of parameters in total energy function and taking derivations with respect to $\hat{\mathbf{U}}$ and $\hat{\mathbf{V}}$ leads to the following coupled electromechanical system of equations for the considered 5LMSSP:

$$\begin{bmatrix} \mathbf{K}_{uu} & \mathbf{K}_{uv} \\ \mathbf{K}_{vu} & -\mathbf{K}_{vv} \end{bmatrix} \begin{Bmatrix} \hat{\mathbf{U}} \\ \hat{\mathbf{V}} \end{Bmatrix} = \begin{Bmatrix} \mathbf{f}_u \\ \mathbf{f}_v \end{Bmatrix} \tag{1.36}$$

where $\mathbf{K}_{uu}$, $\mathbf{K}_{vu} = \mathbf{K}_{uv}^T$, and $\mathbf{K}_{vv}$ are the mechanical, coupling electromechanical and piezo-electric permittivity stiffness matrices of 5LMSSP, respectively. These stiffness matrices and force vectors are defined as:

$$\mathbf{K}_{uu} = \int_A [\mathbf{H}_0^T \ \mathbf{H}_1^T \ \mathbf{H}_3^T] \ \overline{\mathbf{Q}_b} [\mathbf{H}_0 \ \mathbf{H}_1 \ \mathbf{H}_3]^T dA + \int_A [\mathbf{H}_s^T \ \mathbf{H}_s^T] \ \overline{\mathbf{Q}_s} [\mathbf{H}_s \ \mathbf{H}_s]^T dA \quad (1.37)$$

$$\mathbf{K}_{uv} = \int_A [\mathbf{H}_0^T \ \mathbf{H}_1^T \ \mathbf{H}_3^T] \ \overline{\mathbf{E}_{bv}} \mathbf{H}_v dA + \int_A [\mathbf{H}_s^T \ \mathbf{H}_s^T] \ \overline{\mathbf{E}_{sv}} \mathbf{H}_v dA \quad (1.38)$$

$$\mathbf{K}_{vv} = \int_A \mathbf{H}_v^T \overline{\mathbf{k}} \mathbf{H}_v \ dA \quad (1.39)$$

$$\mathbf{f}_u = \int_A \mathbf{H}_w^T f \ dA, \ \mathbf{f}_v = \int_A \mathbf{H}_v^T f_v \ dA \quad (1.40)$$

where:

$$\overline{\mathbf{Q}_b} = \int_{-h/2}^{h/2} \begin{bmatrix} 1 & z & c_1 z^3 \\ & z^2 & c_1 z^4 \\ \text{Sym.} & & c_1^2 z^6 \end{bmatrix} \mathbf{Q}_b dz, \ \overline{\mathbf{Q}_s} = \int_{-h/2}^{h/2} \begin{bmatrix} 1 & 3c_1 z^2 \\ 3c_1 z^2 & 9c_1^2 z^4 \end{bmatrix} \mathbf{Q}_s dz \quad (1.41)$$

$$\overline{\mathbf{E}_{bv}} = \int_{-h/2}^{h/2} \{1 \ z \ c_1 z^3\}^T \mathbf{e}_b \ dz, \ \overline{\mathbf{E}_{sv}} = \int_{-h/2}^{h/2} \{1 \ 3c_1 z^2\}^T \mathbf{e}_s \ dz \quad (1.42)$$

$$\overline{\mathbf{k}} = \int_{-h/2}^{h/2} \mathbf{k} \ dz \quad (1.43)$$

$$\mathbf{H}_w = [0 \ 0 \ \psi_i \ 0 \ 0] \quad (1.44)$$

1.3.3 Essential Boundary Conditions (Mechanical Supports)

Since MLS shape functions are not able to meet the Kronecker delta conditions, the implementation of essential boundary condition needs further effort in comparison with FEMs. In this chapter, the transformation approach is utilized for the implementation of such boundary condition to prevent the increase of computational costs due to applying penalty parameters like the element free Galerkin method [23]. This approach is based on transformation matrices which link the real and generalized nodal values as:

$$\mathbf{U} = \mathbf{R}_u \hat{\mathbf{U}}, \mathbf{V} = \mathbf{R}_v \hat{\mathbf{V}} \quad (1.45)$$

where

$$\mathbf{R}_u = \begin{bmatrix} \psi_1(x_1)\mathbf{I} & \psi_1(x_2)\mathbf{I} & . & . & \psi_1(x_N)\mathbf{I} \\ \psi_2(x_1)\mathbf{I} & \psi_2(x_2)\mathbf{I} & . & . & \psi_2(x_N)\mathbf{I} \\ . & . & . & . & . \\ . & . & . & . & . \\ \psi_N(x_1)\mathbf{I} & \psi_N(x_2)\mathbf{I} & . & . & \psi_N(x_N)\mathbf{I} \end{bmatrix}_{5N\times 5N}, \mathbf{R}_v = \begin{bmatrix} \psi_1(x_1) & \psi_1(x_2) & . & . & \psi_1(x_N) \\ \psi_2(x_1) & \psi_2(x_2) & . & . & \psi_2(x_N) \\ . & . & . & . & . \\ . & . & . & . & . \\ \psi_N(x_1) & \psi_N(x_2) & . & . & \psi_N(x_N) \end{bmatrix}_{N\times N}$$

$$(1.46)$$

where $\mathbf{I}$ and N are the 5×5 identity matrix and the total number of nodes dispersed in the domain, respectively. The substitution of Eq. (1.45) in Eq. (1.36) rearranges the coupled electromechanical system of equations based on the real nodal values as:

$$
\begin{bmatrix} \hat{\mathbf{K}}_{uu} & \hat{\mathbf{K}}_{uv} \\ \hat{\mathbf{K}}_{vu} & -\hat{\mathbf{K}}_{vv} \end{bmatrix} \begin{Bmatrix} \mathbf{U} \\ \mathbf{V} \end{Bmatrix} = \begin{Bmatrix} \hat{\mathbf{f}}_u \\ \hat{\mathbf{f}}_v \end{Bmatrix}
\tag{1.47}
$$

in which

$$
\hat{\mathbf{K}}_{uu} = \mathbf{R}_u^{-T}\mathbf{K}_{uu}\mathbf{R}_u^{-1}, \hat{\mathbf{K}}_{uv} = \mathbf{R}_u^{-T}\mathbf{K}_{uv}\mathbf{R}_v^{-1}, \hat{\mathbf{K}}_{vu} = \mathbf{R}_v^{-T}\mathbf{K}_{vu}\mathbf{R}_u^{-1}, \hat{\mathbf{K}}_{vv} = \mathbf{R}_v^{-T}\mathbf{K}_{vv}\mathbf{R}_v^{-1}
$$
$$
\hat{\mathbf{f}}_u = \mathbf{R}_u^{-T}\mathbf{f}_u, \hat{\mathbf{f}}_v = \mathbf{R}_v^{-T}\mathbf{f}_v
\tag{1.48}
$$

Now, essential boundary conditions can be treated as easy as FEM.

1.4 Numerical Results

In this section, after establishing the efficiency of the mesh-free solution, the effects of layer arrangement and other significant parameters on the electromechanical performance of the considered 5LMSSPs are investigated. In the modeling of 5LMSSP, the material of polymeric parts (core and the matrix of nanocomposite layers) are considered to be epoxy. The enhancement component of the nanocomposite layers is GPLs with length $a^g = 2.5\,\mu m$, width $b^g = 1.5\,\mu m$, and thickness $h^g = 1.5\,nm$ [39]. In addition, PZT-G1195N is the utilized material of piezoceramic layers. The material properties of the utilized material are given in Table 1.1 [18, 39].

1.4.1 Validation

Due to the novelty and originality of this topic, there exists no comparable experimental, numerical or theoretical data for the considered five-layer sandwich plate. Hence, the verification of the developed numerical solution is assessed by the comparison of the electromechanical deflections of smart three-layer plates obtained from this developed solution with those obtained from two different FEMs presented in [18, 31]. The considered three-layer plates are made up by attaching two PZT-G1195N piezoceramic layers on the faces of isotropic metal (Ti-6Al-4V) and ceramic (aluminum oxide) plates. These plates are assumed to be cantilevered and in square shape with isotropic plate thickness $h_c = 5\,mm$, piezoceramic thickness $h_f = 0.1\,mm$, and length $a = 0.4\,m$ under mechanical load $f_0 = 0.1\,kPa$ and three different voltages of $V_0 = 0$, 20, and 40 V. The comparison of tip deflections between the results obtained from the developed mesh-free solutions based

Table 1.1 The material properties of the utilized materials in 5LMSSP.

	E (GPa)	v	ρ (kg/m³)	e_{31} (C/m²)	e_{32} (C/m²)	k_{33} (×10⁻⁹ F/m)
PZT-G1195N [18]	63	0.3	7750	22.86	22.86	15
GPL [39]	1010	0.186	1060	–	–	–
Epoxy [39]	3	0.34	1200	–	–	–

Table 1.2 Electromechanical tip deflections (mm) of smart three-layer plates made up by attaching PZT-G1195N layers on the faces of isotropic Ti-6Al-4V and aluminum oxide plates.

	Aluminum oxide				Ti-6Al-4V			
V_0 (V)	Present FSDT	Present HSDT	[31]	[18]	Present FSDT	Present HSDT	[31]	[18]
0	−0.0893	−0.0898	−0.0895	−0.0895	−0.2543	−0.2556	−0.2544	−0.2546
20	−0.0460	−0.0466	−0.0461	−0.0461	−0.1334	−0.1349	−0.1333	−0.1335
40	−0.0027	−0.0033	−0.0027	−0.0027	−0.0125	−0.0142	−0.0123	−0.0123

on the first- (first-order shear deformation theory, FSDT) and third-order theories with the deflections reported in [18, 31] is given in Table 1.2. It is worth mentioning that FSDT formulation can be obtained from the developed solution by setting $c_1 = 0$ in Eq. (1.12) and applying shear correction factor in Eq. (1.19). The presented results in Table 1.2 show very good agreements between the results which verify the reliability of the reported results in this work.

The reliability of the presented solution is also examined by checking the convergence of the obtained results. Therefore, a simply supported square 5LMSSP with the layer arrangement of case I under the static electromechanical of $P_0 = 1$ kPa and $V_0 = 80$ V is considered. The other specifications of 5LMSSP are considered as $a = 0.5$ m, $h_p = 0.5$ mm, $h_n = 1$ mm, $h_c = 12$ mm, $e_0 = 0.7$, $V_r^* = 0.1$, and $p = 1$. Figure 1.4 illustrates that the central deflection values of the considered 5LMSSP are converged by increasing the number of dispersed nodes in the domain such that increasing node numbers after dispersing 21×21 shows insignificant difference in the values of central deflections.

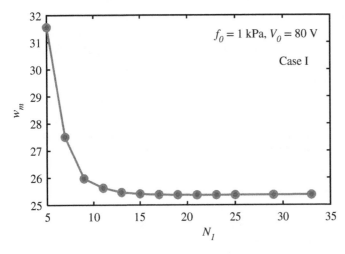

Figure 1.4 Electromechanical deflections of 5LMSSP at center point versus dispersed node numbers along x and y directions ($N_1 \times N_1$).

1.4.2   Static Deflections in 5LMSSPs

This section compares the electromechanical performance of the two case layer arrangements of 5LMSSPs under static mechanical and electrical loads. In the following simulations of this chapter, square simply supported 5LMSSPs with $a = 0.5$ m, $h_p = 0.5$ mm, $h_n = 1$ mm, $h_c = 12$ mm, $e_0 = 0.7$, $V_r^* = 0.1$, and $p = 1$ are considered unless otherwise stated.

Electromechanical loading effect on the center line deflection profile of the considered 5LMSSPs are illustrated in Figure 1.5. This figure shows that the deflection shapes and values of both cases under pure mechanical load ($V_0 = 0$) are almost the same. However, deflection shapes under electromechanical loads show that electrical voltage has a considerable impact on the deformed shape of 5LMSSPs. The case I layer arrangement is more sensitive to electrical voltage than case II. This is because the space between piezoceramic layers in case I is farther than case II which can intensify the impact of electrical load on the deflection of 5LMSSPs. In case I, Figure 1.5a illustrates that applying the electrical voltage of $V_0 = 40$ V completely recovers the deflections due to the mechanical pressure of $f_0 = 0.1$ kPa and makes the 5LMSSP flat. However, Figure 1.5b shows that the voltage of $V_0 = 40$ V is unable to entirely recover the deflection caused by $f_0 = 0.1$ kPa in case II of 5LMSSP such that there are still downward deflections. In addition, the voltage of $V_0 = 40$ V causes reverse deflections with the maximum values around $w_m = 25$ μm and $w_m = 18$ μm at the center points of the considered 5LMSSPs under $f_0 = 0.1$ kPa with layer arrangements of cases I and II, respectively. Overall, Figure 1.5 shows that the layer arrangement of this specific 5LMSSP has a significant impact on electromechanical performance, but an insignificant impact on mechanical performance. However, these impacts can be different by changing the other specifications of 5LMSSP such as the dispersion and volume fraction of graphene, porosity volume, and layer thicknesses.

To examine the impacts of layer arrangement and GPL volume fraction on electromechanical performance, 5LMSSPs in both layer arrangements with uniform dispersion of GPLs (i.e. $p = 0$ or $V_r(z) = V_r^*$) and different V_r^* have been considered. Figure 1.6 shows the central deflections of such smart plates versus the volume fraction of GPLs under pure mechanical load and subjected to electromechanical loads. Generally, the variation of deflections shows that the increase of GPL volume fraction results in the reduction of both mechanical and electromechanical deflections due to the enhancement of the mechanical stiffness of 5LMSSPs. In both cases, the increase of GPL volume fraction has the same impact on the electromechanical performance of 5LMSSPs although the electromechanical deflections of these plates in case I are higher than those of case II. For 5LMSSPs under pure mechanical load, the deflections of case II are higher than those of case I when $V_r^* < 0.5$. However, the deflection induced by pure mechanical loads in both cases of 5LMSSPs are almost the same when GPL volume fraction is $V_r^* > 0.5$ because the Young's modulus of the resulting nanocomposite becomes closer to piezoceramic.

The dispersion of GPLs is another parameter and its impact on the performance of the considered 5LMSSPs with both layer arrangements is shown in Figure 1.7. This figure shows central deflections of the considered plates under mechanical and electromechani-

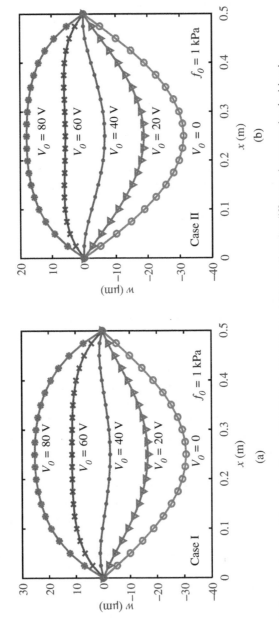

Figure 1.5 Deflection shapes of the center lines of 5LMSSPs in (a) case I and (b) case II under different electromechanical loads.

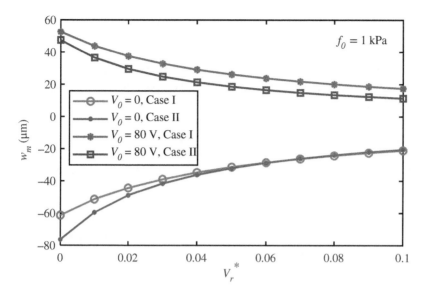

Figure 1.6 Mechanical and electromechanical deflections of 5LMSSPs with the two considered layer arrangements at the center point versus GPL volume fraction.

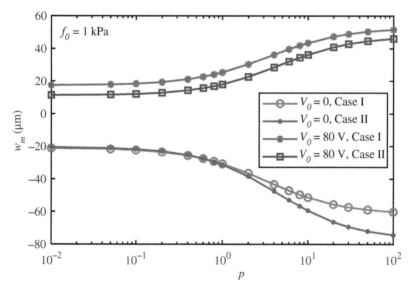

Figure 1.7 Mechanical and electromechanical deflections of 5LMSSPs with the two considered layer arrangements at the center point versus the exponent of GPL volume fraction.

cal loads versus the exponent of GPL volume fraction. Deflection results disclose that the increase of dispersion exponent leads to the increase of deflections due to the reduction of GPL contents in nanocomposite layers. As observed in Figure 1.6, there is a specific exponent value ($p > 1$) that differentiates the pure mechanical responses of such smart plates with layer arrangement of cases I and II. This difference is more evident as the exponent value increases. In addition, there is a constant difference between the electromechanical deflections of 5LMSSPs with the two different layer arrangements as exponent value changes.

The impacts of porosity content on the mechanical and electromechanical performances of the considered smart plates are illustrated in Figure 1.8a,b, respectively. Generally, it is observed that the increase of porosity content does not have a considerable impact on the deflections of 5LMSSPs in both cases. This observation shows that the weight of the considered smart plates can be remarkably reduced without losing its structural bending stiffness. As shown in Figure 1.8a, layer arrangement also has an insignificant impact on the mechanical bending responses of 5LMSSP. However, the electromechanical deflections of 5LMSSP with layer arrangement case I are much higher than those of case II which is due to different spaces between piezoceramic layers in cases I and II.

In the design of such smart plates, layer thicknesses are important parameters. Figure 1.9 shows the deflection variations of 5LMSSPs with layer arrangement cases I and II under mechanical and electromechanical loads for different core thicknesses from $h_c = 5$ to 40 mm. For the considered 5LMSSPs under pure mechanical loads, dramatic reductions are observed by increasing core thickness from $h_c = 5$ mm up to 15 mm in both cases I and II. However, the increase of core thickness to $h_c > 25$ mm does not have a considerable impact on the mechanical deflection of either case of 5LMSSPs. It is also observed that the change of layer arrangement does not have a remarkable impact on the deflections of 5LMSSPs under pure mechanical loads. Moreover, electromechanical deflections disclose that in 5LMSSPs with thin core ($h_c > 10$ mm), layer arrangement has a noticeable impact on deflections such that the contribution of electrical inputs in the obtained deflections of case I is higher than those of case II. With the increase of core thickness, the impact of layer arrangement gradually becomes insignificant because the deflections of both cases are very close when $h_c > 25$ mm.

Figure 1.10 shows the deflection variations of 5LMSSPs with layer arrangement cases I and II under mechanical and electromechanical loads for nanocomposite layers with different thicknesses from $h_n = 0.5$ to 10 mm. Figure 1.11 also illustrates the same deflections of 5LMSSPs versus the thickness change of piezoceramic layers from $h_p = 0.1$ to 1.6 mm. Since nanocomposite layers have higher Young's modulus and farther distances from the neutral axis of 5LMSSP, the investigated variation range of h_n is smaller than that of h_c in the previous modeling of 5LMSSP. In addition, because of the same reasons and the brittle structures of piezoceramic layers, such small values and variation range are considered for h_p. Again, both figures show that layer arrangement has insignificant effect on the deflections induced by pure mechanical loads because the Young's modulus of the considered GPLs-reinforced nanocomposite with $V_r^* = 0.1$ and $p = 1$ is close to that of PZT-G1195N. In addition, 5LMSSPs with layer arrangement case I have higher deflections induced by electromechanical loads in comparison with those of case II. Figure 1.10 shows that the difference between electromechanical deflections is almost constant by increasing h_n, but Figure 1.11 shows that this difference is reduced by increasing h_p because, according to Eq. (1.22), the use of thicker piezoceramic layers reduces

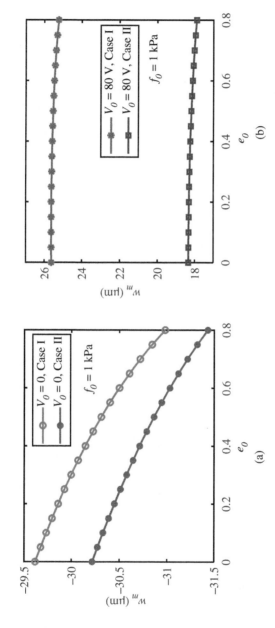

Figure 1.8 (a) Mechanical and (b) electromechanical deflections of 5LMSSPs with the two considered layer arrangements at the center point versus porosity volume.

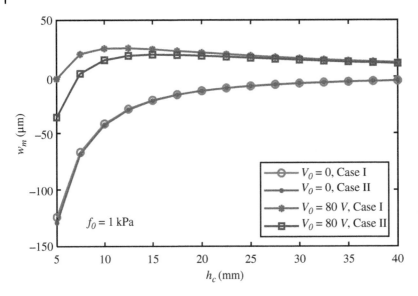

Figure 1.9 Mechanical and electromechanical deflections of 5LMSSPs with the two considered layer arrangements at the center point versus the thickness of core.

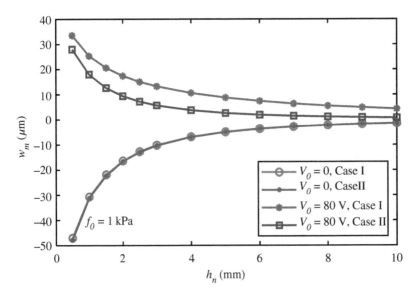

Figure 1.10 Mechanical and electromechanical deflections of 5LMSSPs with the two considered layer arrangements at the center point versus the thickness of nanocomposite layers.

potential differences and consequently the impact of electrical input on the deflections. Overall, the increase of the thickness of either nanocomposite or piezoceramic layers reduces both mechanical and electromechanical deflection of 5LMSSPs with both layer arrangements.

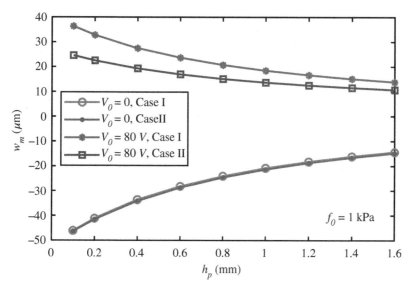

Figure 1.11 Mechanical and electromechanical deflections of 5LMSSPs with the two considered layer arrangements at the center point versus the thickness of piezoceramic layers.

1.5 Conclusions

This work mainly studied the impact of layer arrangement on the electromechanical static deflections of a 5LMSSP. The considered smart plate was made of three different materials including one thick porous polymeric layer as core, two piezoceramic layers as active components, and two FG GPL-reinforced nanocomposite layers. In this work, the impact of considering piezoceramic layers as the faces or middle layers of 5LMSSP on the electromechanical bending responses were examined using Reddy's TSDT and mesh-free solution. The obtained results of 5LMSSPs disclosed that:

- The use of piezoceramic layers as outer layers of 5LMSSPs provides higher electromechanical deflections than that of 5LMSSPs with middle piezoceramic layers.
- The pure mechanical deflections of 5LMSSPs are not significantly affected by changing layer arrangement, especially when GPL volume fraction is above $V_r^* = 0.5$.
- The increase of GPL volume fraction results in the reduction of both mechanical and electromechanical deflections.
- The increase of porosity content does not have a considerable impact on deflections although embedding porosity remarkably reduces structural weight.

References

1 Chopra, I. (2002). Review of state of art of smart structures and integrated systems. *AIAA J.* 40: 2145–2187.

2 Vinson, J.R. (2001). Sandwich structures. *Appl. Mech. Rev.* 54: 201.

3 Birman, V. and Kardomateas, G.A. (2018). Review of current trends in research and applications of sandwich structures. *Compos. Part B Eng.* 142: 221–240.

4 Safaei, B., Moradi-Dastjerdi, R., Qin, Z. et al. (2019). Determination of thermoelastic stress wave propagation in nanocomposite sandwich plates reinforced by clusters of carbon nanotubes. *J. Sandwich Struct. Mater.* https://doi.org/10.1177/109963621984828.

5 Safaei, B., Moradi-Dastjerdi, R., Behdinan, K. et al. (2019). Thermoelastic behavior of sandwich plates with porous polymeric core and CNT clusters/polymer nanocomposite layers. *Compos. Struct.* 226: 111209.

6 Bisht, A., Dasgupta, K., and Lahiri, D. (2018). Effect of graphene and CNT reinforcement on mechanical and thermomechanical behavior of epoxy – a comparative study. *J. Appl. Polym. Sci.* 135: 46101.

7 Singh, N.P., Gupta, V.K., and Singh, A.P. (2019). Graphene and carbon nanotube reinforced epoxy nanocomposites: a review. *Polymer* 180: 121724.

8 Behdinan, K., Moradi-Dastjerdi, R., Safaei, B. et al. (2020). Graphene and CNT impact on heat transfer response of nanocomposite cylinders. *Nanotechnol. Rev.* 9: 41–52.

9 Tao, J., He, X., Yi, S., and Deng, Y. (2019). Broadband energy harvesting by using bistable FG-CNTRC plate with integrated piezoelectric layers. *Smart Mater. Struct.* 28: 095021.

10 Kundalwal, S.I. and Ray, M.C. (2016). Smart damping of fuzzy fiber reinforced composite plates using 1-3 piezoelectric composites. *J. Vib. Control* 22: 1526–1546.

11 Moradi-Dastjerdi, R., Meguid, S.A., and Rashahmadi, S. (2019). Dynamic behavior of novel nanocomposite diaphragm in piezoelectrically-actuated micropump. *Smart Mater. Struct.* 28: 105022.

12 Moradi-Dastjerdi, R., Meguid, S.A., and Rashahmadi, S. (2019). Dynamic behavior of novel micro fuel pump using zinc oxide nanocomposite diaphragm. *Sens. Actuators, A* 297: 111528.

13 Yan, J., Zhou, W., Zhang, X., and Lin, Y. (2019). Interface monitoring of steel-concrete-steel sandwich structures using piezoelectric transducers. *Nucl. Eng. Technol.* 51: 1132–1141.

14 Askari Farsangi, M.A., Saidi, A.R., and Batra, R.C. (2013). Analytical solution for free vibrations of moderately thick hybrid piezoelectric laminated plates. *J. Sound Vib.* 332: 5981–5998.

15 Duc, N.D. and Cong, H.P. (2018). Nonlinear thermo-mechanical dynamic analysis and vibration of higher order shear deformable piezoelectric functionally graded material sandwich plates resting on elastic foundations. *J. Sandwich Struct. Mater.* 20: 191–218.

16 Phung-Van, P., Tran, L.V., Ferreira, A.J.M. et al. (2017). Nonlinear transient isogeometric analysis of smart piezoelectric functionally graded material plates based on generalized shear deformation theory under thermo-electro-mechanical loads. *Nonlinear Dyn.* 87: 879–894.

17 Duc, N.D. (2018). Nonlinear thermo-electro-mechanical dynamic response of shear deformable piezoelectric sigmoid functionally graded sandwich circular cylindrical shells on elastic foundations. *J. Sandw. Struct. Mater.* 20: 351–378.

18 Nguyen-Quang, K., Dang-Trung, H., Ho-Huu, V. et al. (2017). Analysis and control of FGM plates integrated with piezoelectric sensors and actuators using cell-based smoothed discrete shear gap method. *Compos. Struct.* 165: 115–129.

19 Dinh Khoa, N., Thi Thiem, H., and Duc, N.D. (2019). Nonlinear buckling and postbuckling of imperfect piezoelectric S-FGM circular cylindrical shells with metal-ceramic-metal layers in thermal environment using Reddy's third-order shear deformation shell theory. *Mech. Adv. Mater. Struct.* 26: 248–259.

20 Talebitooti, R., Daneshjoo, K., and Jafari, S.A.M. (2016). Optimal control of laminated plate integrated with piezoelectric sensor and actuator considering TSDT and meshfree method. *Eur. J. Mech .A/Solids* 55: 199–211.

21 Phung-Van, P., Nguyen-Thoi, T., Le-Dinh, T., and Nguyen-Xuan, H. (2013). Static and free vibration analyses and dynamic control of composite plates integrated with piezo-electric sensors and actuators by the cell-based smoothed discrete shear gap method (CS-FEM-DSG3). *Smart Mater. Struct.* 22: 095026.

22 Mohammadimehr, M., Zarei, H.B., Parakandeh, A., and Arani, A.G. (2017). Vibration analysis of double-bonded sandwich microplates with nanocomposite facesheets reinforced by symmetric and un-symmetric distributions of nanotubes under multi physical fields. *Struct. Eng. Mech.* 64: 361–379.

23 Moradi-Dastjerdi, R., Meguid, S.A., and Rashahmadi, S. (2019). Electro-dynamic analysis of smart nanoclay-reinforced plates with integrated piezoelectric layers. *Appl. Math. Model.* 75: 267–278.

24 Arani, A.G., Jafari, G.S., and Kolahchi, R. (2018). Vibration analysis of nanocomposite microplates integrated with sensor and actuator layers using surface SSDPT. *Polym. Compos.* 39: 1936–1949.

25 Malekzadeh, P., Setoodeh, A.R., and Shojaee, M. (2018). Vibration of FG-GPLs eccentric annular plates embedded in piezoelectric layers using a transformed differential quadrature method. *Comput. Meth. Appl. Mech. Eng.* 340: 451–479.

26 Jabbari, M., Joubaneh, E.F., Khorshidvand, A.R., and Eslami, M.R. (2013). Buckling analysis of porous circular plate with piezoelectric actuator layers under uniform radial compression. *Int. J. Mech. Sci.* 70: 50–56.

27 Jabbari, M., Hashemitaheri, M., Mojahedin, A., and Eslami, M.R. (2014). Thermal buckling analysis of functionally graded thin circular plate made of saturated porous materials. *J. Therm. Stress* 37: 202–220.

28 Askari, M., Saidi, A.R., and Rezaei, A.S. (2018). An investigation over the effect of piezoelectricity and porosity distribution on natural frequencies of porous smart plates. *J. Sandw. Struct. Mater.* https://doi.org/10.1177/1099636218791092.

29 Barati, M.R. and Zenkour, A.M. (2018). Electro-thermoelastic vibration of plates made of porous functionally graded piezoelectric materials under various boundary conditions. *J. Vib. Control* 24: 1910–1926.

30 Mohammadi, M., Bamdad, M., Alambeigi, K. et al. (2019). Electro-elastic response of cylindrical sandwich pressure vessels with porous core and piezoelectric face-sheets. *Compos. Struct.* 225: 111119.

31 Nguyen, L.B., Nguyen, N.V., Thai, C.H. et al. (2019). An isogeometric Bézier finite element analysis for piezoelectric FG porous plates reinforced by graphene platelets. *Compos. Struct.* 214: 227–245.

32 Nguyen, N.V., Lee, J., and Nguyen-Xuan, H. (2019). Active vibration control of GPLs-reinforced FG metal foam plates with piezoelectric sensor and actuator layers. *Compos. Part B Eng.* 172: 769–784.

33 Setoodeh, A.R., Shojaee, M., and Malekzadeh, P. (2019). Vibrational behavior of doubly curved smart sandwich shells with FG-CNTRC face sheets and FG porous core. *Compos. Part B Eng.* 165: 798–822.

34 Moradi-Dastjerdi, R. and Behdinan, K. (2020). Temperature effect on free vibration response of a smart multifunctional sandwich plate. *J. Sandw. Struct. Mater.* https://doi .org/10.1177/1099636220908707.

35 Moradi-Dastjerdi, R. and Behdinan, K. (2020). Stability analysis of multifunctional smart sandwich plates with graphene nanocomposite and porous layers. *Int. J. Mech. Sci.* 167: 105283.

36 Moradi-Dastjerdi, R., Radhi, A., and Behdinan, K. (2020). Damped dynamic behavior of an advanced piezoelectric sandwich plate. *Compos. Struct.* 243: 112243.

37 Behdinan, K. and Moradi-Dastjerdi, R. (2019). Electro-mechanical behavior of smart sandwich plates with porous core and graphene-reinforced nanocomposite layers. In: *Proceedings of the ASME 2019 International Mechanical Engineering Congress and Exposition*, Salt Lake City, UT, USA (11–14 November 2019), 1–7. ASME.

38 Roberts, A.P. and Garboczi, E.J. (2001). Elastic moduli of model random three-dimensional closed-cell cellular solids. *Acta Mater.* 49: 189–197.

39 Song, M., Kitipornchai, S., and Yang, J. (2017). Free and forced vibrations of functionally graded polymer composite plates reinforced with graphene nanoplatelets. *Compos. Struct.* 159: 579–588.

40 Rafiee, M.A., Rafiee, J., Wang, Z. et al. (2009). Enhanced mechanical properties of nanocomposites at low graphene content. *ACS Nano.* 3: 3884–3890.

41 Reddy, J.N. (2004). *Mechanics of Laminated Composite Plates and Shells: Theory and Analysis*. CRC Press.

42 Lancaster, P. and Salkauskas, K. (1981). Surface generated by moving least squares methods. *Math. Comput.* 37: 141–158.

2

Heat Transfer Behavior of Graphene-Reinforced Nanocomposite Sandwich Cylinders

Kamran Behdinan and Rasool Moradi-Dastjerdi

Advanced Research Laboratory for Multifunctional Lightweight Structures (ARL-MLS), Department of Mechanical & Industrial Engineering, University of Toronto, Toronto, Canada

2.1 Introduction

Graphene is a two-dimensional (2D) material with the thickness of one single atom. This material is a carbon allotrope with nanostructure lattice of hexagonally arranged carbon atoms [1]. Due to the particular shape of graphene, it has an extraordinary thermal conductivity which has been reported to be up to 3000 W/(m·K) [2, 3] and 5300 W/(m·K) [4] while thermal conductivities of polymers are usually less than 1 W/(m·K). This huge difference between thermal conductivities of graphene and polymers, introduces graphene as a highly efficient filler to significantly enhance the thermal conductivity of polymers [4, 5]. In the calculation of thermal conductivity of such nanocomposite materials, agglomeration formation and polymer–graphene interfacial thermal resistance are two significant parameters which can restrict the improvement of thermal behavior [6, 7]. However, the dispersion of nanofillers based on functionally graded (FG) patterns in the host matrix usually improves the overall thermal and mechanical performances of nanocomposite materials. Moreover, FG dispersions of nanofillers provide a better management on the thermomechanical responses of nanocomposite structures [8–13].

   Before conducting thermal analyses on engineering structures, thermal properties including thermal conductivity and specific heat conduction of each component of these structures need to be determined. In nanocomposites, such characterizations are more important due to the existence of unknown scale effects which significantly affect the governing micromechanical models of ordinary composites. In a pioneering work, Balandin et al. [4] considered a single-layer graphene and experimentally measured its thermal conductivity to be in the range of 4840–5300 W/(m·K). Hu et al. [14] performed molecular dynamics (MD) simulations on the thermal conductivity of graphene/organic polymer nanocomposite materials and reported that the negative effect of interfacial thermal resistance could be restricted by the use of graphene with high aspect ratios of up to several thousands. They also showed that the good dispersion of long and thin graphene sheets has a positive effect on the thermal conductivity of such nanocomposite materials. Chu et al. [15] proposed an analytical model to calculate the thermal conductivity of

Advanced Multifunctional Lightweight Aerostructures: Design, Development, and Implementation,
First Edition. Kamran Behdinan and Rasool Moradi-Dastjerdi.

graphene/polymer nanocomposite materials with consideration of interface thermal resistance between components. Im and Kim [16] experimentally and analytically studied the thermal conductivity of a hybrid multi-wall carbon nanotube (CNT)/graphene oxide/epoxy nanocomposite material. Gu et al. [17] also experimentally showed that the addition of 21.4% graphene platelet as filler into polyethylene (PE) increases the thermal conductivity of pure PE by up to nine times. Shen et al. [18] conducted MD simulations to investigate the effect of size and number of graphene layers on the thermal conductivity of multilayer graphene/epoxy nanocomposites. They showed that the increase in number and lateral size of graphene layers significantly improve the thermal conductivity of nanocomposite such that the addition of 2.8% volume fraction of graphite nanoplatelets with more than 10 layers and diameters of larger than 30 μm improve the thermal conductivity of pure epoxy by up to seven times. In addition, aggregation state and the presence of any type of defect in the filler phase are two additional important parameters on the thermal conductivity of nanocomposites. Evans et al. [6] developed a Monte Carlo model to consider the effect of filler aggregation on the thermal conductivity of composites. They found that the enhancement of thermal conductivity is mainly governed by interfacial thermal resistance, the morphology of agglomeration and the conductivity of the filler. Zabihi and Araghi [7] analytically investigated the impacts of CNT and graphene defects including vacancy, Stone–Wales and doped on the effective thermal conductivity of nanocomposites reinforced with graphene, CNT, or a combination of CNT/graphene. Li et al. [19] also studied the effect of defects on the thermal conductivity of nanocomposite materials reinforced with defected graphene using MD simulations. Moheimani et al. [20] developed a closed-form unit cell micromechanical model to study the thermal conductivity of unidirectional CNTs/polymer nanocomposites.

The use of engineering structures made of nanocomposites reinforced with allotropes of carbon (mainly graphene and CNT) has attracted great attention due to the astonishing thermomechanical properties of such nanofillers. Researchers have mostly focused on the thermoelastic responses of nanocomposite structures. Thermoelastic static [21], vibrations [22] and dynamic [23] responses of nanocomposite cylinders reinforced with wavy CNT subjected to thermal and mechanical loads have been investigated. In these works, first, the steady state heat transient behaviors have been analyzed, and then the obtained temperature profiles have been treated as thermal load on nanocomposite cylinders. Alibeigloo [24] evaluated the thermoelastic responses of cylindrical panels reinforced with straight CNTs using an analytical method based on three-dimensional (3D) theory of elasticity. He developed a coupled thermomechanical solution for this structure. Pourasghar et al. [25, 26] employed a generalized differential quadrature (DQ) method to present a 3D response for the thermoelastic analysis of cylinders and cylindrical panels reinforced with CNTs. Safaei et al. conducted one-dimensional (1D) steady state [27] and transient [28] heat transfer analyses across the thickness of sandwich plates with CNT reinforced faces and isotropic polymer core. In these works, the main focus has been on the thermoelastic static and stress wave propagation responses of such advanced sandwich plates. Damadam et al. [29] conducted thermal ratcheting in FG metal/ceramic cylinders subjected to cyclic thermal gradient effect and internal pressure. They considered nonlinear kinematic hardening and presented Bree's diagram of temperature–pressure to optimize the design of such FG cylinders. Pourasghar and Chen [30] developed a solution based on Newton–Raphson and DQ methods to present nonlinear hyperbolic transient heat transfer responses of CNT reinforced nanocomposite cylindrical panels. They also utilized the same method to study (nanoscale) size effect on the transient heat transfer and vibrations of FG nanocomposite microbeams reinforced with CNTs [31]. Jalali et al. [32] also studied the effects of environment temperature and size on the vibrations of FG ceramic/metal Timoshenko beam using the Rayleigh–Ritz method based on modified couple stress theory. They found that temperature effect on the vibrations of FG beams is more noticeable than that on homogeneous ones. Zargar et al. [33] utilized an analytical method to study the steady state temperature profile of circular porous fins with internal heat generation and rectangular cross-section. Using a meshless method, both steady state and transient heat transfer behaviors of axisymmetric nanocomposite cylinders reinforced with CNT were studied in [34]. Moradi-Dastjerdi and Behdinan [35] used a steady state temperature profile of axisymmetric nanocomposite cylinders reinforced with graphene sheets to present thermoelastic static and vibrations of such nanocomposite cylinders.

Due to the specific thermal and mechanical benefits of sandwich structures, they have attracted great attention from both academic researchers and industrial engineers. However, their heat transfer behaviors have not yet been thoroughly investigated. In this work, an axisymmetric sandwich cylinder construction consisting of a homogeneous polymeric layer bounded between two inner and outer FG nanocomposite layers reinforced with graphene sheets has been considered. Transient and steady state heat transfer behaviors of such sandwich cylinders under various types of thermal boundary conditions including thermal flux, convection environment and constant temperature have been studied using a mesh-free method. Moving least square (MLS) shape functions and the transformation approach have been utilized in the proposed mesh-free method. In addition, a micromechanical model has been employed to calculate the thermal conductivity of nanocomposite layers. In this model, graphene sheets are assumed to be randomly oriented.

2.2   Modeling of Sandwich Cylinders

The schematic diagram of the proposed nanocomposite sandwich cylinder under various thermal boundary conditions including heat flux, constant temperature, and convection environment with convective heat transfer coefficient h_a and ambient temperature T_a is shown in Figure 2.1. The inner and outer layers of the sandwich cylinder are assumed to be made of nanocomposite reinforced with randomly oriented graphene sheets. The length, inner radius, outer radius, cylinder thickness, and nanocomposite layer thickness are shown with L, r_i, r_o, H, and h, respectively.

2.2.1   Dispersion of Graphene Sheets

Taking the concept of FG ceramic/metal materials, the dispersion of graphene sheets can follow smooth variation profiles to optimize the use of such nanofillers and end up with the best design for engineering structures made of polymer/graphene. In these regards, the dispersion of graphene sheets in the inner and outer nanocomposite layers is considered to

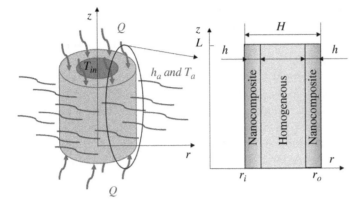

Figure 2.1 Axisymmetric nanocomposite sandwich cylinders reinforced with graphene sheets.

have smooth FG patterns along the radial direction. The FG patterns are mathematically represented as [36]:

$$
\begin{cases}
f_r = f_r^{max}\left(1 - \dfrac{r}{h} + \dfrac{r_i}{h}\right)^{\lambda} & r_i < r < r_i + h \\
f_r = 0 & r_i + h < r < r_o - h \\
f_r = f_r^{max}\left(1 + \dfrac{r}{h} - \dfrac{r_o}{h}\right)^{\lambda} & r_o - h < r < r_o
\end{cases}
\tag{2.1}
$$

where f_r is the volume fraction of graphene sheets which changes from zero to its highest value f_r^{max}. The variation of f_r is managed with volume fraction exponent λ. Figure 2.2 illustrates the variation of graphene volume fraction for different values of λ. According to this figure, the overall volume fraction of graphene is increased as λ decreases. In addition, $\lambda = 1$ represents a linear pattern of graphene dispersion.

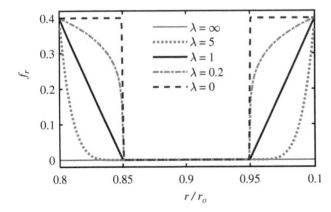

Figure 2.2 FG patterns of graphene distribution along the radial direction of the sandwich cylinder.

2.2.2 Thermal Properties

There are different models for the estimation of the thermal conductivity of nanocomposite materials. Based on Monte Carlo simulations, Foygel et al. [37] proposed a nonlinear equation between the thermal conductivity and nanofiller volume fraction. Moreover, Deng et al. [38] suggested a simple model for the thermal conductivity of nanocomposite materials which considers the effects of different nanofiller aspects including aspect ratio, volume fraction, anisotropic property, thermal conductivity and interfacial thermal resistance. Zabihi and Araghi [7] combined these two models and proposed the following equation for the thermal conductivity of nanocomposite materials based on graphene aspect ratio p, thermal conductivity k^r and volume fractions f_r as [6, 7]:

$$\frac{k}{k^m} = 1 + \frac{F_c}{H + \left(\frac{K^f}{k^m} - 1\right)} \tag{2.2}$$

where

$$H = \frac{1}{p^2 - 1}\left[\frac{p}{\sqrt{p^2 - 1}}\ln(p + \sqrt{p^2 - 1}) - 1\right] \tag{2.3}$$

$$F_c = \frac{2}{3}\left(f_r - \frac{1}{p}\right)^{1.5}, K^f = \frac{k^r}{1 + (2R_k k^r/l)} \tag{2.4}$$

where l, k^m, and R_k are graphene length, the thermal conductivity of polymer, and the thermal resistance of graphene–polymer interface, respectively.

Moreover, using a simple rule of mixture, the specific heat conduction c_p and density ρ of graphene/polymer nanocomposites can be roughly determined as follows:

$$c_p = f_r c_p^r + f_m c_p^m, \rho = f_r \rho^r + f_m \rho^m \tag{2.5}$$

where $f_m = 1 - f_r$ is polymer volume fraction, and the superscripts m and r are utilized to imply polymer and graphene, respectively.

2.2.3 Governing Thermal Equations

Due to the presence of axisymmetric thermal loading and sandwich cylinder, the governing equation is reduced from 3D to 2D. For FG axisymmetric sandwich cylinders, governing steady state and transient thermal equations are determined as [34, 39]:

$$\frac{1}{r}\left[k(r)\frac{\partial}{\partial r}\left(r\frac{\partial T(r, z)}{\partial r}\right)\right] + k(r)\frac{\partial^2 T(r, z)}{\partial z^2} = 0 \tag{2.6}$$

$$\frac{1}{r}\left[k(r)\frac{\partial}{\partial r}\left(r\frac{\partial T(r, z)}{\partial r}\right)\right] + k(r)\frac{\partial^2 T(r, z)}{\partial z^2} = \rho(r)c_p(r)\frac{\partial T(r, z)}{\partial t} \tag{2.7}$$

where the temperature at point $\mathbf{X} = \mathbf{X}_i(r, z)$ is shown by $T(r, z)$ and time is shown by t. In addition, boundary and initial conditions are defined as:

$$T = T_1 \text{ on } \Gamma_1, Q = Q_2 \text{ on } \Gamma_2, h = h_a \text{ and } T_a \text{ on } \Gamma_3$$

$$T(r, z) = T^0 \text{ at } t = t_0 \tag{2.8}$$

where Γ shows all faces of the sandwich cylinder. It should be mentioned that only one boundary condition can be applied on each boundary of the sandwich cylinder.

2.3 Mesh-Free Formulations

MLS shape function is one of the most famous shape functions in mesh-free methods. Using this shape function, exact temperature domain **T** can be estimated in some predefined nodes. However, this shape function does not satisfy the Kronecker delta property. Therefore, MLS shape function Φ estimates the generalized values of temperature domain $\hat{\mathbf{T}}$ in its support domain as follows [40]:

$$\hat{\mathbf{T}} = \sum_{i=1}^{n} \Phi_i \, T_i = \Phi\,\mathbf{T} \tag{2.9}$$

where

$$\mathbf{T} = [T_1, T_2, .., T_n]^T, \hat{\mathbf{T}} = [\hat{T}_1, \hat{T}_2, .., \hat{T}_n]^T, \Phi = [\Phi_1, \Phi_2, .., \Phi_n]^T \tag{2.10}$$

where n is the number of nodes located inside the support domain of the shape function. Furthermore, MLS shape function at $\mathbf{X} = \mathbf{X}_i(r, z)$ is determined as [40]:

$$\Phi_i(\mathbf{X}) = \underbrace{\mathbf{P}^T(\mathbf{X})[\mathbf{M}(\mathbf{X})]^{-1}w(\mathbf{X} - \mathbf{X}_i)\mathbf{P}(\mathbf{X}_i)}_{(1\times1)} \tag{2.11}$$

where w is the cubic spline weight function, **P** is the base vector and **M** is the moment matrix. **P** and **M** in axisymmetric problems are determined as:

$$\mathbf{P}(\mathbf{X}) = [1, r, z]^T, \mathbf{M}(\mathbf{X}) = \left[\sum_{i=1}^{n} w(\mathbf{X} - \mathbf{X}_i)\mathbf{P}(\mathbf{X}_i)\mathbf{P}^T(\mathbf{X}_i) \right] \tag{2.12}$$

Taking the weak forms of Eqs. (2.6) and (2.7) and using MLS shape functions as interpolation and weight functions (applying the Galerkin procedure), the mesh-free forms of steady state and transient thermal governing equations are derived as [34]:

$$\mathbf{K}_T\hat{\mathbf{T}} - \int_{\Gamma} \Phi^T(-M_T\mathbf{T} + S)d\Gamma = 0 \tag{2.13}$$

$$\mathbf{C}\frac{d\hat{\mathbf{T}}}{dt} + \mathbf{K}_T\hat{\mathbf{T}} - \int_{\Gamma} \Phi^T(-M_T\mathbf{T} + S)d\Gamma = 0 \tag{2.14}$$

for boundaries subjected to thermal flux, $M_T = 0$ and $S = Q$. Also, for boundaries subjected to a convection environment, $M_T = h$ and $S = hT_a$. By the imposition of these two types of thermal boundary conditions, Eqs. (2.13) and (2.14) are regenerated as follows:

$$\mathbf{K}_T\hat{\mathbf{T}} + \mathbf{K}_M\hat{\mathbf{T}} = \mathbf{Q} \tag{2.15}$$

$$\mathbf{C}\frac{d\hat{\mathbf{T}}}{dt} + \mathbf{K}_T\hat{\mathbf{T}} + \mathbf{K}_M\hat{\mathbf{T}} = \mathbf{Q} \tag{2.16}$$

where

$$\mathbf{K}_T = \int_{\Omega} \mathbf{B}^T \mathbf{D}\,\mathbf{B}\, dv, \mathbf{K}_M = \int_{\Gamma} M_T\Phi^T\Phi dv,$$

$$\mathbf{C} = \int_{\Omega} (\rho c_p)\Phi^T\Phi dv, \mathbf{Q} = \int_{\Gamma} Q\Phi^T ds \tag{2.17}$$

in which

$$
\mathbf{B} = \begin{bmatrix} \dfrac{\partial \Phi_1}{\partial r} & \dfrac{\partial \Phi_2}{\partial r} & \cdots & \dfrac{\partial \Phi_n}{\partial r} \\[2ex] \dfrac{\partial \Phi_1}{\partial z} & \dfrac{\partial \Phi_2}{\partial z} & \cdots & \dfrac{\partial \Phi_n}{\partial z} \end{bmatrix}, \mathbf{D} = k(r)\begin{bmatrix} 1 & 0 \\ 0 & 1 \end{bmatrix}
\tag{2.18}
$$

It should be noted that Eqs. (2.14) and (2.15) are generated based on the generalized values of temperature domain $\widehat{\mathbf{T}}$, and therefore, the essential boundary conditions (boundaries with known temperature) cannot be directly applied to these two equations. This shortcoming arises because MLS shape function does not satisfy the Kronecker delta property, as mentioned earlier. In this regard, transformation matrix $\mathbf{R}$ is employed to bridge the generalized values of domain to real values of calculated domain $\overline{\mathbf{T}}$ as follows [34]:

$$
\overline{\mathbf{T}} = \mathbf{R}\,\widehat{\mathbf{T}}
\tag{2.19}
$$

where the transformation matrix is described as:

$$
\mathbf{R} = \begin{bmatrix} \Phi_1(\mathbf{X}_1) & \Phi_2(\mathbf{X}_1) & \cdot & \cdot & \Phi_n(\mathbf{X}_1) \\ \Phi_1(\mathbf{X}_2) & \Phi_2(\mathbf{X}_2) & & & \Phi_n(\mathbf{X}_2) \\ \cdot & \cdot & \cdot & \cdot & \cdot \\ \cdot & \cdot & \cdot & \cdot & \cdot \\ \Phi_1(\mathbf{X}_n) & \Phi_2(\mathbf{X}_n) & \cdot & \cdot & \Phi_n(\mathbf{X}_n) \end{bmatrix}
\tag{2.20}
$$

Introducing Eq. (2.19) in Eqs. (2.15) and (2.16) results in:

$$
\widehat{\mathbf{K}}_T\overline{\mathbf{T}} + \widehat{\mathbf{K}}_M\overline{\mathbf{T}} = \widehat{\mathbf{Q}}
\tag{2.21}
$$

$$
\widehat{\mathbf{C}}\frac{d\overline{\mathbf{T}}}{dt} + \widehat{\mathbf{K}}_T\overline{\mathbf{T}} + \widehat{\mathbf{K}}_M\overline{\mathbf{T}} = \widehat{\mathbf{Q}}
\tag{2.22}
$$

where

$$
\begin{aligned}
\widehat{\mathbf{C}} &= \mathbf{R}^{-T}{\cdot}\mathbf{C}{\cdot}\mathbf{R}^{-1}, \widehat{\mathbf{q}} = \mathbf{T}^{-T}{\cdot}\mathbf{q} \\
\widehat{\mathbf{K}}_T &= \mathbf{R}^{-T}{\cdot}\mathbf{K}_T{\cdot}\mathbf{R}^{-1}, \widehat{\mathbf{K}}_M = \mathbf{R}^{-T}{\cdot}\mathbf{K}_M{\cdot}\mathbf{R}^{-1}
\end{aligned}
\tag{2.23}
$$

By the application of this modification, Eqs. (2.21) and (2.22) have been prepared to impose the essential thermal boundary conditions as easy as finite element method (FEM) equations.

2.4 Results and Discussion

For the proposed FG nanocomposite sandwich cylinders, the thermal responses of steady state and transient heat transfer governing equations are presented in this section. Overall, two different materials have been utilized in the sandwich cylinder. Graphene sheets are used as the reinforcement part of nanocomposite layers. In addition, PE is considered as the polymeric part of nanocomposite layers and the middle homogeneous layer. The material characterization of the utilized materials [7, 17] is as follows:

PE: $k^m = 0.5$ W/(m $\cdot$ K), $c_p^m = 1466$ J/(kg$\cdot$K), $\rho^m = 1150$ kg/m^3

Graphene: $k^r = 3000$ W/(m $\cdot$ K), $c_p^r = 700$ J/(kg$\cdot$K), $\rho^r = 4118$ kg/m^3, $l = 40$ μm, $p = 250$, $R_k = 3.5 \times 10^{-9}$

2.4.1 Thermal Conductivity of Graphene/PE Nanocomposite

The use of Eq. (2.2) results in the thermal conductivity of graphene/PE nanocomposite as a function of graphene volume fraction. Figure 2.3 shows such variation and compares k predicted by Eq. (2.2) with the experimental values reported in [17]. The results show the significant increase of thermal conductivity with the increase of graphene volume fraction such that the use of 25% volume fraction of graphene in PE results in improved k of the resulting nanocomposite around 10 times that of k^m (the thermal conductivity of pure PE). Moreover, a good agreement with experimental results is observed for the predicted thermal conductivities.

2.4.2 Verification

The temperature profile in axisymmetric long cylinders (1D heat conduction) with constant thermal conductivity is determined using the following equation reported by Hetnarski and Eslami [39]:

$$T = \frac{T_i - T_o}{\ln(r_o - r_i)} \ln(r_o - r) + T_b \tag{2.24}$$

where T_i and T_o are the temperatures at the inner and outer radii. To verify the accuracy of the developed mesh-free method, the temperature profile predicted by the mesh-free method is compared with the temperature profile evaluated from Eq. (2.20) (exact solution) in Figure 2.4 for a cylinder with $T_i = 320$ K and $T_o = 290$ K. It can be seen that the mesh-free method provides excellent accuracy in calculation of the steady state temperature profile of axisymmetric cylinders.

In addition, the convergence of the developed mesh-free method is verified for the transient heat transfer response of a nanocomposite sandwich cylinder. In this regard, the thermal responses of steady state and transient heat transfer equations are solved for the considered cylinder. It is expected that temperature profiles of transient response are gradually converged to the temperature profile obtained from steady state solution. The considered sandwich cylinder is assumed to be subjected to $T_i = 320$ K, $T_o = 290$ K and thermally insulated lower and upper faces (i.e. $Q_d = 0$ and $Q_u = 0$). Its initial temperature is $T^0 = 300$ K

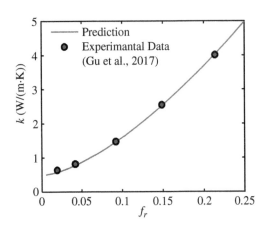

Figure 2.3 Thermal conductivity of graphene/PE nanocomposite versus the volume fraction of graphene.

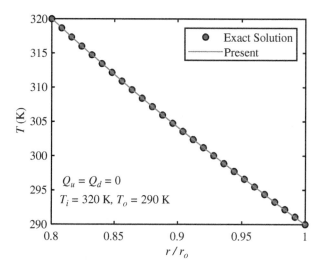

Figure 2.4 Verification of temperature profile along a cylinder thickness evaluated by the mesh-free method in comparison with exact solution. Source: Based on [39].

and graphene volume fraction linearly ($\lambda = 1$) varies along the radial direction according to Eq. (2.1) when $f_r^{\max} = 0.4$. Moreover, the dimensions of sandwich cylinders are $r_i = 80$ mm, $r_o = 100$ m, $h_f = 0.25\,H$, and $L = 40$ mm. Figure 2.5 confirms that the expected convergence happened such that after 500 s, the temperature profile of transient solution is identical to the one obtained from steady state solution.

2.4.3 Heat Transfer Response

In this subsection, the heat transfer behavior of the proposed nanocomposite sandwich cylinder is investigated to study the effects of graphene content, cylinder thickness, and thermal boundary conditions.

In the first simulation, sandwich cylinders with $r_i = 80$ mm, $r_o = 100$ m, $h_f = 0.25\,H$, and $L = 40$ mm subjected to $T_i = 320$ K, $T_o = 290$ K, $Q_d = 0$, and $Q_u = 0$ are considered to study the effect of graphene volume fraction on their thermal responses. Two linear ($\lambda = 1$) FG sandwich cylinders with $f_r^{\max} = 0.1$ and $f_r^{\max} = 0.4$ are considered and their thermal responses are compared with a uniform nanocomposite sandwich cylinder with $f_r = 0.4$ and a homogeneous polymeric cylinder (i.e. $f_r = 0$). Figure 2.6a illustrates the time histories of the considered cylinders at their midpoints ($r = r_{\text{mid}}$ and $z = L/2$). The shortest ($t = 200$ s) and longest ($t = 800$ s) stationary times belong to sandwich cylinders with $f_r = 0.4$ and $f_r = 0$, respectively. This observation clearly shows that the use of graphene and the increase of its content result in a significant reduction in the stationary time of such cylinders. The reason is that the overall thermal conductivity of these cylinders is improved as graphene content increases. Figure 2.6b shows the steady state temperature profiles of the considered sandwich cylinders at $z = L/2$ and $t = 800$ s. The figure shows that by increasing thermal conductivity of nanocomposite layers, the increase of graphene content leads to a better heat or coldness transfer from the nearest boundary. Moreover, due to constant thermal conductivity along the thickness, a smooth temperature profile is observed in the homogeneous

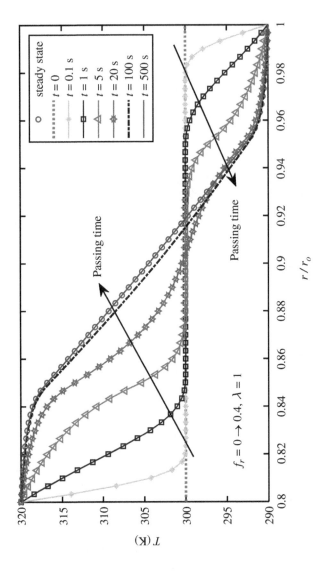

Figure 2.5 The convergence of transient temperature profiles to steady state temperature profile in a graphene/PE nanocomposite sandwich cylinder with $T_i = 320\,K$, $T_o = 290\,K$, $Q_d = 0$, and $Q_u = 0$.

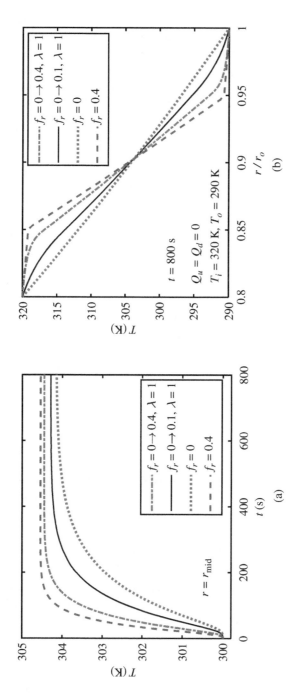

Figure 2.6 The effect of graphene volume fraction on the (a) time histories of temperature at midpoints and (b) steady state profiles of temperature at $z = L/2$ in nanocomposite sandwich cylinders with $r_i = 80$ mm.

polymeric cylinder. For the same reason, smooth temperature profiles happen in all the layers of the sandwich cylinder with $f_r = 0.4$.

In the simulation, the previous sandwich cylinders with lower thickness (i.e. $r_i = 90\,\text{mm}$) are reconsidered. Figure 2.7 shows the time histories of temperature at midpoint (Figure 2.7a) and steady state temperature profiles at $z = L/2$ for these sandwich cylinders (Figure 2.7b). Comparison of Figures 2.6 and 2.7 reveals that the decrease of cylinder thickness significantly accelerates the time needed to reach steady state condition such that in homogeneous cylinders, stationary time is reduced from $t = 800\,\text{s}$ to around $t = 200\,\text{s}$ in the thinner cylinder and in sandwich nanocomposite cylinders with $f_r = 0.4$, it is decreased from $t = 200\,\text{s}$ to around $t = 50\,\text{s}$.

By changing the dispersion pattern of graphene sheets (changing λ) in the cylinders of the first simulation, this effect on stationary time and their final temperature profiles is investigated in Figure 2.8 when $f_r^{\max} = 0.4$. This figure shows that graphene dispersion in nanocomposite layers has a remarkable effect on the transient and steady state heat transfer of such nanocomposite sandwich cylinders such that cylinders with higher λ require longer times to reach their steady state conditions. In addition, heat or coldness transfer is weaker in cylinders with higher λ due to the presence of a lower amount of graphene, as shown in Figure 2.2.

Finally, Figure 2.9 illustrates the temperature profiles of nanocomposite sandwich cylinders subjected to $h_o = h_u = 100\,\text{W/(m}^2 \cdot \text{K)}$, $T_{a-o} = T_{a-u} = 250\,\text{K}$, $Q_i = 1\,\text{kW/m}^2$, and $Q_d = 2\,\text{kW/m}^2$ for different graphene dispersions at their steady state conditions. The results show that the decrease of λ improves heat transfer, such that the inner temperature of cylinder with $\lambda = 0.2$ is 280 K and this cylinder has the lowest overall temperature. Moreover, the highest temperature gradient in both radial and axial directions is observed in nanocomposite sandwich cylinders with $\lambda = 5$.

2.5 Conclusions

In this work, the transient and steady state heat transfer behaviors of sandwich cylinders with graphene/polymer nanocomposite layers subjected to different thermal boundary conditions were studied through an axisymmetric model. A micromechanical model was employed to estimate the thermal properties of nanocomposite materials. In addition, the governing equations of heat transfer were treated by a mesh-free method. Studying the effects of thermal boundary conditions and graphene content on the heat transfer behavior of nanocomposite sandwich cylinders resulted in the following main conclusions:

- Due to the improvement of thermal conductivity, the use of graphene and the increase of its content result in a significant reduction in the stationary time of such cylinders.
- The increase of graphene content leads to a better heat or coldness transfer from the nearest boundary.
- Graphene dispersion in nanocomposite layers has a remarkable effect on transient and steady state heat transfer.
- The decrease of cylinder thickness significantly accelerates the time needed to reach steady state condition such that in one case, the stationary time is reduced from $t = 200\,\text{s}$ to around $t = 50\,\text{s}$ in the thinner cylinder.

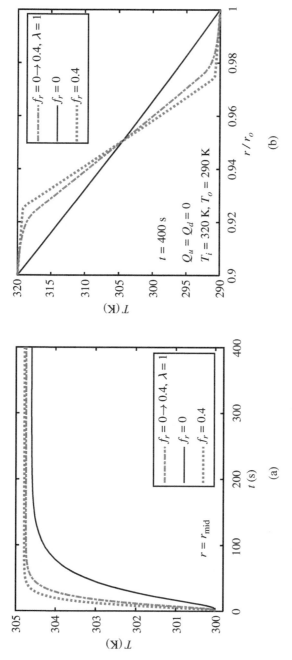

Figure 2.7 The effect of graphene volume fraction on the (a) time histories of temperature at midpoints and (b) steady state profiles of temperature at $z = L/2$ in nanocomposite sandwich cylinders with $r_i = 90$ mm.

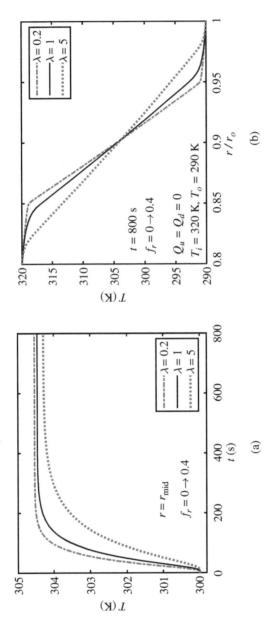

Figure 2.8 The effect of graphene dispersion on the (a) time histories of temperature at midpoints and (b) steady state profiles of temperature at $z = L/2$ in nanocomposite sandwich cylinders with $r_i = 80$ mm.

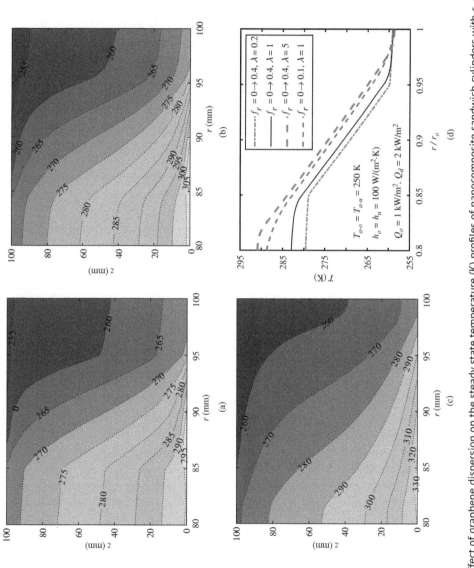

Figure 2.9 The effect of graphene dispersion on the steady state temperature (K) profiles of nanocomposite sandwich cylinders with $r_i = 80$ mm: (a) $f_r = 0 \rightarrow 0.4$, $\lambda = 0.2$; (b) $f_r = 0 \rightarrow 0.4$, $\lambda = 1$; (c) $f_r = 0 \rightarrow 0.4$, $\lambda = 5$; and (d) at $z = L/2$.

References

1 Pop, E., Varshney, V., and Roy, A.K. (2012). Thermal properties of graphene: fundamentals and applications. *MRS Bull.* 37: 1273–1281.

2 Fujii, M., Zhang, X., Xie, H. et al. (2005). Measuring the thermal conductivity of a single carbon nanotube. *Phys. Rev. Lett.* https://doi.org/10.1103/PhysRevLett.95.065502.

3 Samani, M.K., Khosravian, N., Chen, G.C.K. et al. (2012). Thermal conductivity of individual multiwalled carbon nanotubes. *Int. J. Therm. Sci.* 62: 40–43.

4 Balandin, A.A., Ghosh, S., Bao, W. et al. (2008). Superior thermal conductivity of single-layer graphene. *Nano Lett.* 8: 902–907.

5 Balandin, A.A. (2011). Thermal properties of graphene and nanostructured carbon materials. *Nat. Mater.* 10: 569–581.

6 Evans, W., Prasher, R., Fish, J. et al. (2008). Effect of aggregation and interfacial thermal resistance on thermal conductivity of nanocomposites and colloidal nanofluids. *Int. J. Heat Mass Transf.* 51: 1431–1438.

7 Zabihi, Z. and Araghi, H. (2017). Effective thermal conductivity of carbon nanostructure based polyethylene nanocomposite: influence of defected, doped, and hybrid filler. *Int. J. Therm. Sci.* 120: 185–189.

8 Moradi-Dastjerdi, R. and Behdinan, K. (2020). Stability analysis of multifunctional smart sandwich plates with graphene nanocomposite and porous layers. *Int. J. Mech. Sci.* 167: 105283.

9 Qin, Z., Zhao, S., Pang, X. et al. (2020). A unified solution for vibration analysis of laminated functionally graded shallow shells reinforced by graphene with general boundary conditions. *Int. J. Mech. Sci.* 170: 105341.

10 Qin, Z., Safaei, B., Pang, X., and Chu, F. (2019). Traveling wave analysis of rotating functionally graded graphene platelet reinforced nanocomposite cylindrical shells with general boundary conditions. *Results Phys.* 15: 102752.

11 Shokri-Oojghaz, R., Moradi-Dastjerdi, R., Mohammadi, H., and Behdinan, K. (2019). Stress distributions in nanocomposite sandwich cylinders reinforced by aggregated carbon nanotube. *Polym. Compos.* 40: E1918–E1927.

12 Safaei, B., Moradi-Dastjerdi, R., Behdinan, K., and Chu, F. (2019). Critical buckling temperature and force in porous sandwich plates with CNT-reinforced nanocomposite layers. *Aerosp. Sci. Technol.* 91: 175–185.

13 Safaei, B. and Fattahi, A.M. (2017). Free vibrational response of single-layered graphene sheets embedded in an elastic matrix using different nonlocal plate models. *Mechanika* 23: 678–687.

14 Hu, L., Desai, T., and Keblinski, P. (2011). Thermal transport in graphene-based nanocomposite. *J. Appl. Phys.* 110: 1–6.

15 Chu, K., Jia, C., and Li, W. (2012). Effective thermal conductivity of graphene-based composites. *Appl. Phys. Lett.* 101: 121916.

16 Im, H. and Kim, J. (2012). Thermal conductivity of a graphene oxide-carbon nanotube hybrid/epoxy composite. *Carbon N Y* 50: 5429–5440.

17 Gu, J., Li, N., Tian, L. et al. (2015). High thermal conductivity graphite nanoplatelet/UHMWPE nanocomposites. *RSC Adv.* 5: 36334–36339.

18 Shen, X., Wang, Z., Wu, Y. et al. (2016). Multilayer graphene enables higher efficiency in improving thermal conductivities of graphene/epoxy composites. *Nano Lett.* 16: 3585–3593.

19 Li, M., Zhou, H., Zhang, Y. et al. (2018). Effect of defects on thermal conductivity of graphene/epoxy nanocomposites. *Carbon N Y* 130: 295–303.

20 Moheimani, R. and Hasansade, M. (2019). A closed-form model for estimating the effective thermal conductivities of carbon nanotube–polymer nanocomposites. *Proc. Inst. Mech. Eng. Part C J. Mech. Eng. Sci.* 233: 2909–2919.

21 Moradi-Dastjerdi, R., Payganeh, G., and Tajdari, M. (2018). Thermoelastic analysis of functionally graded cylinders reinforced by wavy CNT using a mesh-free method. *Polym. Compos.* 39: 2190–2201.

22 Moradi-Dastjerdi, R. and Payganeh, G. (2018). Thermoelastic vibration analysis of functionally graded wavy carbon nanotube-reinforced cylinders. *Polym. Compos.* 39: E826–E834.

23 Moradi-Dastjerdi, R. and Payganeh, G. (2017). Thermoelastic dynamic analysis of wavy carbon nanotube reinforced cylinders under thermal loads. *Steel Compos. Struct.* 25: 315–326.

24 Alibeigloo, A. (2016). Elasticity solution of functionally graded carbon nanotube-reinforced composite cylindrical panel subjected to thermo mechanical load. *Compos. Part B Eng.* 87: 214–226.

25 Pourasghar, A., Moradi-Dastjerdi, R., Yas, M.H. et al. (2018). Three-dimensional analysis of carbon nanotube- reinforced cylindrical shells with temperature- dependent properties under thermal environment. *Polym. Compos.* 39: 1161–1171.

26 Pourasghar, A. and Chen, Z. (2016). Thermoelastic response of CNT reinforced cylindrical panel resting on elastic foundation using theory of elasticity. *Compos. Part B Eng.* 99: 436–444.

27 Safaei, B., Moradi-Dastjerdi, R., Behdinan, K. et al. (2019). Thermoelastic behavior of sandwich plates with porous polymeric core and CNT clusters/polymer nanocomposite layers. *Compos. Struct.* 226: 111209.

28 Safaei, B., Moradi-Dastjerdi, R., Qin, Z. et al. (2019). Determination of thermoelastic stress wave propagation in nanocomposite sandwich plates reinforced by clusters of carbon nanotubes. *J. Sandw. Struct. Mater.* https://doi.org/10.1177/109963621984828.

29 Damadam, M., Moheimani, R., and Dalir, H. (2018). Bree's diagram of a functionally graded thick-walled cylinder under thermo-mechanical loading considering nonlinear kinematic hardening. *Case Stud. Therm. Eng.* 12: 644–654.

30 Pourasghar, A. and Chen, Z. (2019). Hyperbolic heat conduction and thermoelastic solution of functionally graded CNT reinforced cylindrical panel subjected to heat pulse. *Int. J. Solids Struct.* 163: 117–129.

31 Pourasghar, A. and Chen, Z. (2019). Effect of hyperbolic heat conduction on the linear and nonlinear vibration of CNT reinforced size-dependent functionally graded microbeams. *Int. J. Eng. Sci.* 137: 57–72.

32 Jalali, M.H., Zargar, O., and Baghani, M. (2018). Size-dependent vibration analysis of FG microbeams in thermal environment based on modified couple stress theory. *Iran J. Sci. Technol. Trans. Mech. Eng.* https://doi.org/10.1007/s40997-018-0193-6.

33 Zargar, O., Mollaghaee-Roozbahani, M., Bashirpour, M., and Baghani, M. (2019). The application of homotopy analysis method to determine the thermal response of convective-radiative porous fins with temperature-dependent properties. *Int. J. Appl. Mech.* https://doi.org/10.1142/s1758825119500881.

34 Moradi-Dastjerdi, R. and Payganeh, G. (2017). Transient heat transfer analysis of functionally graded CNT reinforced cylinders with various boundary conditions. *Steel Compos. Struct.* 24: 359–367.

35 Moradi-Dastjerdi, R. and Behdinan, K. (2019). Thermoelastic static and vibrational behaviors of nanocomposite thick cylinders reinforced with graphene. *Steel Compos. Struct.* 31: 529–539.

36 Jalal, M., Moradi-Dastjerdi, R., and Bidram, M. (2019). Big data in nanocomposites: ONN approach and mesh-free method for functionally graded carbon nanotube-reinforced composites. *J. Comput. Des. Eng.* 6: 209–223.

37 Foygel, M., Morris, R.D., Anez, D. et al. (2005). Theoretical and computational studies of carbon nanotube composites and suspensions: electrical and thermal conductivity. *Phys. Rev. B Condens. Matter Mater Phys.* 71: 1–8.

38 Deng, F., Zheng, Q.S., Wang, L.F., and Nan, C.W. (2007). Effects of anisotropy, aspect ratio, and nonstraightness of carbon nanotubes on thermal conductivity of carbon nanotube composites. *Appl. Phys. Lett.* https://doi.org/10.1063/1.2430914.

39 Hetnarski, R.B. and Eslami, M.R. (2009). *Thermal Stresses–Advanced Theory and Applications*. Springer.

40 Lancaster, P. and Salkauskas, K. (1981). Surface generated by moving least squares methods. *Math Comput.* 37: 141–158.

3

Multiscale Methods for Lightweight Structure and Material Characterization

Vincent Iacobellis and Kamran Behdinan

Advanced Research Laboratory for Multifunctional Lightweight Structures (ARL-MLS), Department of Mechanical & Industrial Engineering, University of Toronto, Toronto, Canada

3.1   Introduction

The development of novel lightweight material and structural designs has been strongly influenced by considering the interaction between multiple length and time scales within a system. One example is the design of materials for crack resistance on the nanoscale such that their evolution at small scales influences the material fracture and fatigue resistance at the macroscale [1–3]. Another example is the design of microscale lattice topologies that are being used as "architected materials" which when looked at from a macroscale perspective represent a homogeneous material with specifically designed properties such as increased strength to weight ratio, cooling, and energy absorption [4–6].

The aforementioned examples could be modeled using solely fine scale numerical methods such as molecular dynamics (MD) or discrete element methods (DEMs) in order to capture the multiscale nature of the problem, however, this would not be computationally feasible, requiring the solution of billions of degrees of freedom for system sizes of any consequence. Multiscale modeling couples multiple numerical methods, which are applicable to different scales, together in one simulation such that the back and forth interaction between the nano to macroscales can be understood and tailored to specific design criteria. The use of multiscale methods forms the basis of the Integrated Computational Materials Engineering (ICME) method increasingly being used in aerospace, automotive and biomedical industries to design, develop, and optimize engineering materials before they are manufactured in a laboratory [7, 8]. This is especially true with the advent and growth of additive manufacturing [9, 10] technologies that allow control in the framework of the underlying structure as well as the ability for rapid prototyping which facilitates the product development process.

In general, there are two categories of multiscale methods: hierarchal and concurrent. Hierarchal approaches take the response at the fine scale and use it to inform the coarse scale continuum properties through a homogenization procedure. In turn the coarse scale response is used to transfer deformation and stress fields to the fine scale. Concurrent multiscale methods "embed" the atomistic model directly into the continuum by

Advanced Multifunctional Lightweight Aerostructures: Design, Development, and Implementation,
First Edition. Kamran Behdinan and Rasool Moradi-Dastjerdi.

decomposing the system into separate atomistic and continuum domains coupled by a transitional domain or boundary.

One concurrent multiscale modeling approach that has been developed, called the bridging cell method (BCM) [11, 12], has been applied to several applications in lightweight material characterization. The BCM is a multiscale method that uses a common finite element framework to solve the atomistic, bridging, and continuum domains. The atomistic and continuum domains are coupled through a bridging domain where both atoms and elements overlap, and atomic displacements are mapped to the finite element nodes through a series of interpolation functions that also aid in defining the weighting of the two energy formulations. A temperature dependent potential is used to introduce temperature effects, thus avoiding the small timesteps associated with MD.

This chapter will review the application of the BCM to multiscale modeling of materials used in lightweight design. The next section will provide a review of hierarchical and concurrent multiscale modeling approaches in the application of lightweight structures. This will be followed by an overview of the BCM along with the extent of its application space, followed by a case study in the integration of BCM with other multiscale approaches to modeling damage in lightweight composite materials.

3.2 Overview of Multiscale Methodologies and Applications

3.2.1 Hierarchical Methods

Hierarchical multiscale methods, also referred to as sequential methods, separate the model into two or more simulations using a step-by-step bottom-up approach. Each individual simulation is performed independently at separate time and length scales. The scales are bridged using various methods such as optimization, statistical analysis, or more commonly homogenization techniques [13].

One such approach is asymptotic homogenization, also known as mathematical homogenization, which is based on asymptotic expansion analysis of periodic structures [14–16]. This approach applies an asymptotic expansion of continuum fields (i.e. displacement, stress, etc.) based on the ratio of the characteristic size of the fine scale heterogeneities and the macrostructure length. Fish et al. [17] developed a generalized asymptotic homogenization procedure based on eigenstrains. It was applied to nonlinear analysis of composite materials [18] as well as damage analysis [19] coupling a heterogeneous continuum microscale model to a continuum macroscale model.

Computational homogenization differs from asymptotic homogenization in that rather than employing a set of governing equations which include both the coarse and fine scale descriptions, it evaluates material information at the fine scale and then passes it to a coarser one in separate solution steps [20]. Thus, these methods compute the stress–strain at points of interest (i.e. finite element integration points) in the coarse scale by solving the detailed model of the fine scale unit cell, or representative volume element (RVE), at that point [20]. The common ground amongst this class of techniques is that they are based on the Hill–Mandel relationship [21] which is a condition that requires the virtual work at a location on the coarse scale to equal the volume average of the virtual work on the fine scale

RVE. One benefit of these methods is that they can be applied using various numerical methods including the Voronoi cell method [22, 23], fast Fourier transforms as proposed by Moulinec and Suquet [24, 25], and most commonly the finite element method [26, 27].

Hierarchical approaches require that the material deforms homogeneously, and difficulties arise when the periodicity of the material is broken by the presence of inhomogeneities such as defects, dislocations, and other failure phenomena. Hierarchical methods also require a clear separation of length scales with separate linear and nonlinear formulations required for most methods. For nonlinear problems the computational time tends to dramatically increase as multiple finite element simulations are being run on each microstructure at each macrostructure integration point.

3.2.2 Concurrent Methods

Atomistic to continuum concurrent multiscale methods decompose a system into an atomistic, a continuum, and a bridging or handshake domain. It is in this handshake domain where atomistic and continuum formulations are coupled to each other. To perform this coupling there are generally two approaches used. The first method is to reduce the spacing of the finite element mesh to the atomic lattice at the atomistic/continuum interface. In this sense there is a direct displacement coupling between the finite element nodes and atoms at the domain interface.

One of the first methods to implement this coupling approach was developed by Tadmor et al. [28–31] and was termed the quasicontinuum (QC) method. In this method, finite element nodes represent groups of atoms whose deformation follow the Cauchy–Born rule [28]. In regions of linear deformation each finite element mesh is less dense and the group of atoms assigned to a node is larger. In regions of nonlinear deformation, the nodes are on a one to one basis with the atoms and the finite element mesh is resolved down to the atomic spacing. More recent advances in the QC method have focused on the extension of the QC method to finite temperature simulations [32, 33]. Similar methods that implement a direct coupling approach include the coupling of length scales method [34], finite element atomistic method [35], and the coupled atomistic discrete dislocation method [36], which allows for the passing of dislocations from the atomistic domain to a discrete dislocation dynamics informed continuum domain.

The second approach to coupling is to have nonlocal and local energy formulations overlap each other and for the coupling to be applied in an average sense. Using this approach, the continuum finite element mesh does not have to be resolved down to the atomic spacing. One such approach is the bridging domain method, developed by Xiao and Belytschko [37–40], which employs a handshake domain where both atomistic and continuum domains overlap coupled together using Lagrange multipliers. Another method developed by Liu et al. [41, 42], called the bridging scale method, applies the coupling using a least squares approach, employing the same philosophy as the bridging domain method where both use a weak coupling but avoid the complex meshing associated with direct coupling.

The above description of existing multiscale methods is not extensive. Many others exist, including the perfectly matched multiscale method [43], heterogeneous multiscale method [44], embedded statistical coupling method [45], and coarse-grained molecular dynamics [46]. For further review on existing concurrent multiscale methods one can refer to [47] as well as [48, 49] for a comparison of existing techniques.

Concurrent multiscale coupling strategies come with their own set of issues when it comes to their implementation and accuracy. One issue is the interaction between the differing energy descriptions in the atomistic and continuum domains. The energy in the continuum domain is local in nature meaning the strain energy at a location is dependent on the deformation at that location alone. The atomistic domain, however, is described by a nonlocal energy formulation where the energy at an atom location is dependent on the deformation of all the atoms within the neighboring vicinity. Since atoms at the continuum interface do not have a full set of neighboring interaction atoms, the result is an inconsistent energy potential resulting in spurious forces acting between the domains.

In addition to the presence of spurious forces within the system there is also the issue of using an MD simulation in the atomistic domain which can lead to the reflection of high frequency phonon waves that begin in the atomistic domain and cannot be transferred into a coarser continuum domain resulting in them being reflected back at the domain interface. Furthermore, the use of MD or other dynamic discrete methods in the atomistic domain require very small timesteps resulting in long simulation times and high strain rates acting on the system. To address this issue, possible solutions have been proposed including multiple timestepping [50, 51] as well as integrating the temperature into the interatomic energy function either directly [52] or through consideration of the vibrational modes of the atomistic system [53, 54].

3.3 Bridging Cell Method

As discussed previously, one alternative concurrent method that looks to address some of the issues discussed in Section 3.2 is the BCM. The BCM was first developed in 2012 [11], being refined over the years to expand its application space. It has been applied to modeling several lightweight material phenomena some examples of which are discussed in Section 3.4 along with its integration into an overall multiscale framework presented in Section 3.5.

The following outlines the formulation for the BCM, however for a more comprehensive description please refer to [11, 55, 56]. Like the concurrent methods discussed, the BCM decomposes the system into an atomistic and continuum domain both of which are then connected through a bridging domain as shown in Figure 3.1.

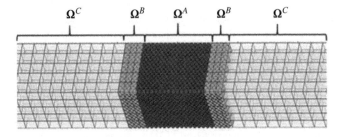

Figure 3.1   BCM model separated into atomistic (Ω^A), bridging (Ω^B), and continuum (Ω^C) domains.

In the bridging domain the constitutive formulation is based on the nonlocal energy potential of the underlying atomic structure with corrections made to this formulation based on weighted contribution from the overlapping continuum elements/cells. Thus, the energy in the bridging domain consists of contributions from both the local continuum energy and the nonlocal atomistic energy. Thus, the overall energy of the system, E_{tot}, is composed of nonlocal and local energy contributions:

$$E_{tot} = E_C^L + E_B^L + E_B^{NL} + E_A^{NL} \tag{3.1}$$

where subscripts A, C, and B refer to the atomistic, continuum, and bridging domains, respectively. Superscripts NL and L refer to nonlocal and local energy formulations, respectively. Thus, the atomistic domain consists of a nonlocal energy contribution, the continuum domain consists of a local energy contribution, and in the bridging domain both nonlocal and local energy contributions exist. Therefore, the energy in the bridging domain can be decomposed into

$$E_B^L = \sum_{c=1}^{n_{cell}} \sum_{q=1}^{n_q} w_q^{(c)} \psi(\mathbf{F}_c) V_c$$

$$E_B^{NL} = \sum_{c=1}^{n_{cell}} \sum_{i=1}^{n_{aB}} U_i \tag{3.2}$$

where ψ is the strain energy density, $\mathbf{F}_c$ is the deformation gradient of cell c, V_c is the volume of cell c, n_{cell} is the number of bridging cells, U_i is the site energy for atom i, and n_{aB} is the number of atoms located inside the bridging domain. The weighting, $w_q^{(c)}$, assigned to the quadrature points q of each cell c is derived from the ratio of the total atomic nonlocal energy contained within the bridging element/cell to the total continuum local energy of that element/cell. This can be represented by,

$$w^{(c)} = 1 - \frac{\frac{1}{2} \mathbf{u}_c^T K_c^{NL} \mathbf{u}_c}{\psi_c(\mathbf{F}_c) \cdot V_c} \tag{3.3}$$

where $\mathbf{u}_c$ are prescribed displacements for cell c, while $\mathbf{K}_c^{NL}$ is the nonlocal stiffness matrix of cell c. By straining each cell and comparing the strain energy based on the nonlocal stiffness to the strain energy of the cell based on a local energy formulation, the weighting between atomistic and continuum energies can be calculated such that the internal forces acting on the cell nodes are zero under conditions of uniform strain. Without such weighting, out-of-balance internal forces would arise withing the bridging domain because of the atoms close to the bridging/continuum domain interface not having a full set of neighboring atoms to interact with. Without a full set of neighboring atoms, the nonlocal energy definition of the atoms will differ from that of an atom within the atomistic domain causing an inconsistent energy definition across the three domains.

To implement the coupling between atomistic and continuum domains, the atom displacements in the bridging domain are mapped to the overlapping elements using a set of interpolation, or basis functions, thus enforcing a strong coupling between element nodes and atoms while avoiding the need to reduce the mesh spacing down to interatomic lengths. The atomistic domain is reformulated into a finite element discretization [57] such that each domain can be solved using a finite element framework based on

$$K\Delta x = F - f \tag{3.4}$$

where

$$K = \frac{\partial^2 E_{tot}}{\partial x \partial x}$$

$$f = -\frac{\partial E_{tot}}{\partial x} \qquad (3.5)$$

where the total energy E_{tot} is defined in Eq. (3.1) to Eq. (3.3), x is the vector of combined node and atom displacements, and F and f are the external and internal forces, respectively. As discussed previously, the coupling in the bridging domain is done through a set of mapping functions, M, which in this case are the element shape functions. High energy atomic configurations can still develop within the bridging domain if deformation of the system becomes too large due to the strict coupling between nodes and atom displacements. When these high energy configurations occur, the degrees of freedom associated with bridging domain atoms are unmapped and allowed to relax freely once the residual in the continuum domain is below a pre-set tolerance. The bridging domain atoms are then remapped and the simulation proceeds. Incorporating the mapping function into Eqs. (3.4) and (3.5), the coupled finite element equation is:

$$[K_{CB}^L + M^T K_{AB}^{NL} M]\Delta x = \{F_{CB}^L + M^T F_{AB}^{NL}\} - \{f_{CB}^L + M^T f_{AB}^{NL}\} \qquad (3.6)$$

where M^T is the transpose matrix of the bridging domain mapping, $K_{CB}{}^L$ is the stiffness matrix corresponding the local energy formulation of the continuum and bridging domains, $K_{AB}{}^{NL}$ is the stiffness matrix corresponding to the nonlocal energy formulation of the atomistic and bridging domains, $F_{CB}{}^L$ and $f_{CB}{}^L$ are applied and internal forces acting on the continuum and bridging domain nodes, respectively, and $F_{AB}{}^{NL}$ and $f_{AB}{}^{NL}$ are applied and internal forces acting on the atomistic and bridging domain atoms, respectively. The derivation of Eq. (3.6) can be found in [11]. The global stiffness matrix can be solved using a nonlinear finite element solver and contains both the nonlocal stiffness interactions from the atomistic and bridging domains and the local stiffness interactions originating from the continuum domain.

To incorporate temperature in the quasistatic solution method, the BCM has employed a temperature dependent potential into the energy formulation. One such method is the local quasiharmonic approximation [53, 54, 56], where the vibrational energy of the atoms about their minimum energy configuration is incorporated into the potential. By solving the system using a temperature dependent potential, the BCM acts to minimize the free energy of the system as opposed to the potential energy. The result is an approach which allows to solve isothermal problems under quasistatic loading conditions. If modeling more dynamic conditions is a necessity, the BCM equations could be solved by incorporating a dynamic finite element solver [58].

3.4 Applications

3.4.1 Crack Propagation in Nickel Single Crystals

In the aerospace industry, where lightweight design is paramount, turbine blades made of nickel-base alloy single crystals are frequently being used due to their excellent strength to weight ratio, material properties at high temperature, and resistance to fatigue and creep. Crack growth often initiates at the nanoscale and the nature of the crack propagation dictates the failure response of the blade macrostructure. Crack path relative to the crystal orientation as well as temperature are two parameters critical in dictating the extent of crack growth and overall life of the turbine blade.

The BCM has been applied to the modeling of crack propagation due to cyclic loading conditions in nickel single crystals [56]. Two crystal orientations were studied; one in which brittle fracture was induced as shown in Figure 3.2a and another where crack tip blunting produced a ductile failure mode. A mode-I crack opening was induced in the system by displacing the ends of the model. The loading was applied cyclically which resulted in the crack propagating with each load cycle.

Crack propagation at 0 (no temperature effect), 400, 650, and 900 K were considered for both brittle and ductile orientations. In all temperature cases the same general dislocation and crack propagation patterns were observed. However, there was significant difference in the crack growth rates where higher temperatures corresponded to greater crack propagation for a given load cycle. From the multiscale simulations the crack growth rate and stress intensity factor were also calculated. These results were plotted, as shown in Figure 3.2b, and the data was fit to the Paris law [59]:

$$\frac{da}{dN} = C(\Delta K)^m \tag{3.7}$$

where da/dN is the crack growth rate, K is the stress intensity factor, and C and m are Paris parameters used for fitting the data. It was found that with increasing temperature the Paris parameter C tended to increase while there was little change in parameter m [56].

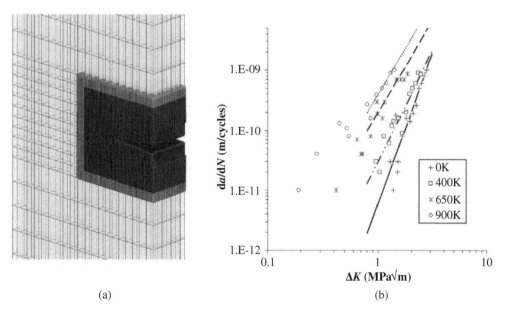

(a)   (b)

Figure 3.2 (a) Crack opening and system deformation for brittle crystal orientation. (b) Crack growth rate versus stress intensity factor for brittle crack orientation with Paris law trend lines for temperatures 0 K (–), 400 K (–···–), 650 K (– –), and 900 K (····).

Furthermore, increasing temperature saw a decrease in the threshold stress intensity factor, indicating that higher temperatures resulted in the onset of crack propagation occurring at a lower stress as well as faster crack propagation once the threshold stress was reached. This type of behavior agreed well with experimental results for single crystal metals [60, 61] and other MD based simulations of crack growth under cyclic loading [62].

3.4.2 Aluminum–Carbon Nanotube Nanocomposite

Metal matrix composites have found a broad range of advanced engineering application especially in the aerospace sector [63, 64]. Metal matrix nanocomposites, and especially those where carbon nanotubes (CNTs) are the main reinforcing agent, experience unique fracture behavior. This fracture behavior is often driven by nonlinear interfacial interactions, load transfer, and CNT breakage mechanisms. In particular, understanding interfacial load and crack propagation from matrix to CNT are critical in implementing these materials for practical use.

Using the BCM, a RVE of an aluminum (Al) matrix CNT reinforced nanocomposite was studied as outlined in [12]. The RVE represents the properties of a single unit cell of the material. This cell is the smallest volume over which a calculation of the material properties is representative of the macroscale properties. In this study, the RVE consists of the CNT and interface region contained within the atomistic domain and the surrounding Al matrix modeled using a linear elastic continuum formulation. Different CNT types were considered including single walled nanotubes (SWNTs), multi-walled nanotubes (MWNTs), and SWNT bundles. The total size of the model was $8.3 \times 8.2 \times 8.1$ nm with a 1 nm diameter CNT filler representing a total composite volume fraction of 4.2%. A constant strain was applied to the RVE in order to obtain the elastic modulus of the system. From these simulations it was found that the MWNT reinforced nanocomposite had the highest modulus at 88.9 GPa while the modulus for the SWNT and SWNT bundle were 84.7 and 87.3 GPa, respectively. For each case, the elastic modulus compared well with both fully atomistic and experimental values [65, 66].

Also studied was the crack propagation through the Al matrix and its interaction with different types of CNT reinforcement as shown in Figure 3.3. The simulation size was $30 \times 5.6 \times 8.2$ nm with an initial precrack applied in the (1 1 1) crystal plane of the Al matrix. The CNT diameter was 1 nm which based on the overall size of the system represented a volume fraction of 5.8%. The system was deformed in the Z-direction (refer to Figure 3.3) to produce type I fracture conditions. The stress versus strain in the system is plotted in Figure 3.4 for MWNT and SWNT type fillers where it can be seen that there is an initial increase in stress to just over 5 GPa followed by a decrease in stress due to failure around the Al matrix. At this stage, the shear stress at the interface between Al matrix and CNT reinforcement is decreasing as the crack propagation induces dislocations and interfacial debonding. Following the drop in stress, the presence of the CNT resists further crack propagation and the CNT itself begins to bear most of the load leading to an increase in the stress. This corresponds to a toughening effect that has been observed experimentally in other nanocomposites and represents one of the potential key benefits of incorporating nanoreinforcements into composite matrices.

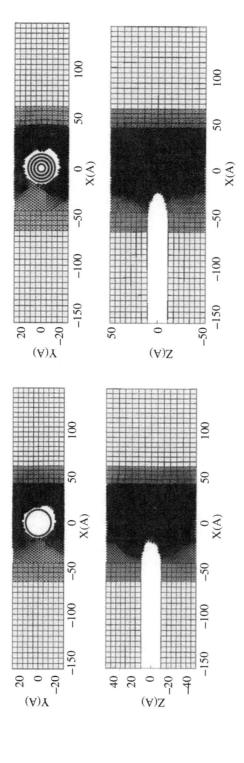

Figure 3.3 Top and side views of BCM model of crack growth in Al−CNT nanocomposite for (a) SWCNT and (b) MWCNT.

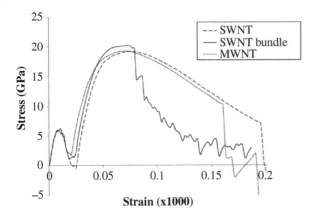

Figure 3.4 Stress–strain relationship for crack propagation in Al–CNT nanocomposite.

3.4.3 Ceramic Composites

Nextel is an important material used in the aeroengine industry due to its high strength-to-weight ratio and its ability to retain its mechanical properties in harsh, high temperature environments. Alumina is one of the main constituents of Nextel and thus being able to accurately simulate the behavior of alumina in the presence of defects is vital in understanding the fracture resistance of Nextel when used in various applications [67–69]. Zamani et al. [69] used the BCM to study the influence of structural nanodefects in the form of nanovoid, pre-crack and magnesium (Mg) impurities on the mechanical response of alumina. Two crystal planes were studied, the (0001) crystal plane (C-plane) and the (1100) plane (M-plane). The system modeled was 100×100 nm with a biaxial strain applied in steps of 0.2% strain. The simulation was run for different defects located at the center of the model for temperatures of 300, 400, 800, 1200, and 1600 K.

Different fracture behavior was observed in the two crystal planes where, unlike the M-plane, the C-plane resisted the propagation of elongated nanocracks and nanovoids in the crystal cleavage plane as shown in Figure 3.5. In Figure 3.5b crack widening around the nanovoid is visible in the cleavage planes of the M-plane, whereas very little crack widening is observed in the C-plane shown in Figure 3.5a. A similar finding is observed for the nanocrack defect shown in Figure 3.6 where again little crack widening was observed in the C-plane shown Figure 3.6a compared with the M-plane shown in Figure 3.6b. Table 3.1 provides a summary of the strength and failure strain for the two crystal planes at different temperatures and in the presence of different defects. The defect free structure was also studied with the tensile strength listed for a strain of 4.5% as an absolute failure strain could not be observed accurately without the presence of a defect in the system. As can be seen from Table 3.1, there was a decrease in ultimate tensile strength with increasing temperature for all defect conditions and crystal orientations. It was also noted that below a temperature of 1200 K, C-plane and M-plane failure response did not change significantly; however, beyond this temperature a transition in failure strain occurred for the elongated crack and the nanovoid defects. For the defect condition of Mg impurities this failure strain transition temperature was observed at 1400 K.

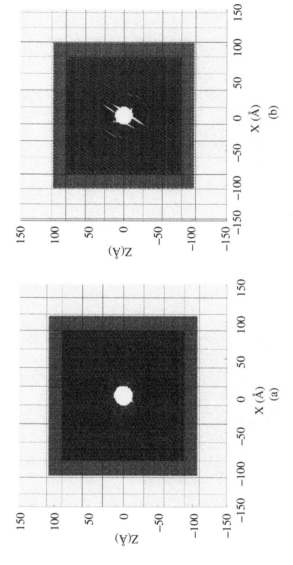

Figure 3.5 BCM model of deformation around nanovoid in alumina for (a) C-plane and (b) M-plane.

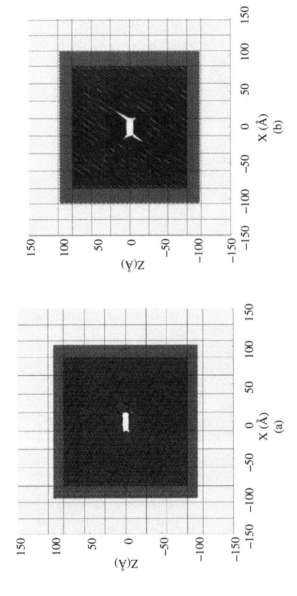

Figure 3.6 BCM model of deformation around nanocrack in alumina for (a) C-plane and (b) M-plane.

Table 3.1 Ultimate tensile strength (UTS) and failure strain for no defect structure and defect structures.

Temperature (K)	Structure	C-plane		M-plane	
		UTS (GPa)	Failure strain (%)	UTS (GPa)	Failure strain (%)
300	No defect	28.59	–	12.04	–
	Nanovoid	15.23	4.5	4.83	4.5
	Nanocrack	9.74	3	7.17	3
	Mg impurities	23.56	6.75	10.96	3.5
1200	No defect	27	–	11.23	–
	Nanovoid	14.53	4	4.44	4
	Nanocrack	9.39	2.75	6.74	2.75
	Mg impurities	22.24	6.75	10.32	3.5
1600	No defect	25.88	–	10.41	–
	Nanovoid	13.56	4	4.08	4
	Nanocrack	8.49	2.75	6.25	2.75
	Mg impurities	21.1	6.5	9.75	3.25

Overall, it was found that for the C-plane, Mg impurities, although reducing the ultimate tensile strength, have less effect on the strength of the material compared with the void and crack defects. For the M-plane, Mg impurities had a significant effect on the failure strain. The most dangerous defect for both crystal planes was the elongated crack, which significantly decreased the ultimate tensile strength through an accelerated brittle fracture. For the M-plane, nanovoids in the presence of biaxial tensile loading was found to have the greatest effect on ultimate tensile strength [69].

3.5 Multiscale Modeling of Lightweight Composites

In Section 3.4, application of BCM in modeling and understanding nano to microscale phenomena was presented. To fully incorporate a multiscale framework from nano to macroscale, a multistep methodology is required. Where BCM has strengths in combining atomistic simulation with continuum micromechanics, it is limited in its application to problems on the structural level as there are limits on the size of the atomistic domain that can be solved for. Thus, the following case study presents how the BCM, or any concurrent based multiscale methodology, can be integrated into an overall multiscale framework.

To demonstrate this multistep framework, the BCM along with cohesive zone micromechanics modeling are used to study the failure behavior in lightweight polymer matrix composites (PMCs). Full details of this approach are outlined in [55, 70]. PMCs are of interest in understanding nonlinear failure mechanisms as they often have set failure modes based on the composition and morphology of the constituent materials. For instance, a limiting design criterion in composites is the transverse strength as high local stresses occur near the fiber–matrix interface inducing debonding and eventual fracture. The incorporation of cohesive zone elements into a finite element model of the composite microstructure in order to capture the fiber–matrix debonding is increasingly being used [71]. Cohesive zone models are often built based on experimental data or best practices, however, to obtain this information is time consuming, expensive, and does not lend itself to rapid development of new composite designs. Thus, it is desirable to have a means of obtaining the cohesive zone parameter from simulation purely based on the physics of the system. In this sense, both a quantitatively useful parameter can be obtained in addition to understanding the underlying mechanism driving the interfacial cohesive interactions.

In this study, the BCM was used to obtain interfacial material properties of a polyimide–graphite interface that were used to build the cohesive zone model constitutive law. The results obtained included the cohesive energy and the cohesive stress across the polyimide–graphite. The cohesive zone model derived from the BCM simulation was used to define a finite cohesive zone model of a unidirectional carbon fiber polyimide matrix composite microstructural RVE. Based on the response of the microstructure RVE to transverse loading conditions, the macroscale strength and modulus were obtained along with their change with respect to temperature [55].

3.5.1 Nano to Microscale: BCM

The polyimide–graphite interface model is shown in Figure 3.7. The atomistic domain was populated with coarse-grained polyimide chains at an initial density of $0.75 \, \text{g/cm}^3$. The carbon fiber surface structure was represented by graphite atomic structure and was contained entirely in the bridging domain as it only underwent linear deformation during the simulation. An MD simulation was run on the system with no external loading to equilibrate the system by removing unnatural high energy configurations created by populating the atomistic domain with the polymer chains. After this equilibration process, the atomistic domain was coupled with a finite element continuum domain for an overall system size $51.4 \times 20.2 \times 125 \, \text{nm}$ with a $2.5 \, \text{nm}$ thick bridging domain.

From the equilibrated structure, the distribution of polymer chains as a function of the distance of the interface was measured as shown in Figure 3.8a. The density distribution of the polymer chains was observed to fluctuate about a mean value of $1.37 \, \text{g/cm}^3$ which corresponded to the bulk density of polyimide. Closer to the interface there was a sharp peak in density corresponding to a build up of polymer chains due to a strong interaction between the graphite and polyimide: an observation also made in past simulations of polymer composite interfaces [72] and experimentally as well [73].

This interfacial interaction is also visible in Figure 3.8b of the average displacement of the polyimide chains relative to the material interface at two different levels of strain. At 5% strain, the displacement profile is relatively linear. This strain state corresponds to the system prior to reaching the point of yielding. However, the displacement profile for 25% strain, corresponding to the post-yield condition, shows a nonlinear relationship with position up until 7 nm where the material again begins to deform linearly. This indicates that the deformation of the polyimide post-yielding is localized to an interfacial zone of finite size close to the interface as was also observed for the density distribution [55]. This localized

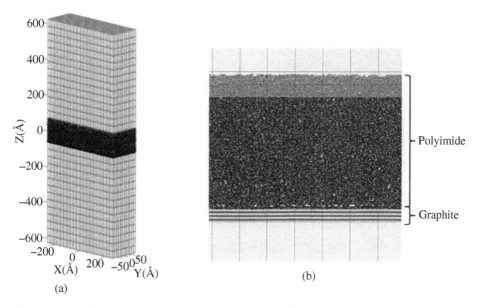

Figure 3.7 (a) BCM model of composite interface and (b) distribution of polymer coarse-grained atoms and graphite atoms in the atomistic and bridging domains.

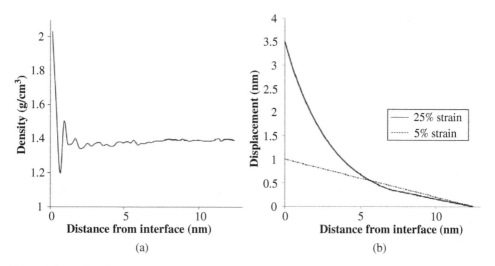

Figure 3.8 (a) Density distribution in the polyimide and (b) displacement distribution of polyimide with respect to the graphite interface.

zone of increased deformation has also been observed experimentally [74] and in full MD simulations [75] where the interfacial zone ranged from 6 to 10 nm in size.

From the tensile deformation of the system, the interfacial stress and fracture energy, or work of adhesion, across the interface of the two materials was calculated. The system deformation produced opening mode failure at the polyimide–graphite interface, as shown in Figure 3.9a. The interfacial stress versus strain was plotted as shown in Figure 3.9b

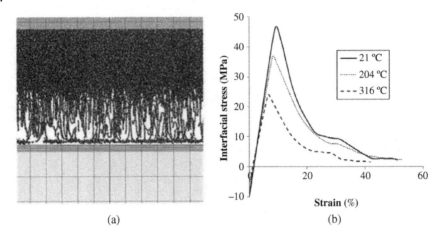

(a)   (b)

Figure 3.9 (a) Separation at the interface and (b) interfacial stress versus strain for tensile loading case.

for temperatures of 21, 204, and 316 °C. Between 21 and 316 °C there was a 45% drop in maximum interfacial stress. In each case, the interfacial stress initially increased linearly as the material faces separated up to maximum stress level. At this point the interfacial stress decreased until leveling off before complete debonding occurred.

3.5.2 Micro to Macroscale: Cohesive Zone Modeling

After getting the fracture energy and interfacial stress from the BCM simulation, the next step was to pass these parameters into a model of the composite microstructure. In this sense, the whole multiscale framework is integrating both the concurrent multiscale modeling approach of the BCM with the hierarchical approach discussed in Section 3.2.1.

The cohesive zone model approximates a traction–displacement relationship along a material interface by being embedded into zero-thickness cohesive zone elements. In this study, a mixed-mode bilinear form of the cohesive zone model is used. The mixed-mode bilinear model incorporates both normal and tangential cohesive tractions, with a region of linear elastic loading followed by a linear softening region. The microstructure was represented by a periodic RVE of the unidirectional fiber reinforced composite as shown in Figure 3.10a. To produce the finite element model of the RVE a random fiber distribution was created with a fiber volume fraction of 56% and fiber diameter of 3 μm. The RVE was deformed in the transverse direction by displacing the top face of the RVE while keeping the bottom face fixed.

From the stress contour plot shown in Figure 3.10b, it was apparent that the microscopic stress distribution is strongly dependent on the arrangement of fibers within the microstructure. This was a result of high localized stresses occurring in regions where the interfiber distance was the shortest with respect to the direction of loading [55]. It was observed that crack initiation due to debonding between fiber and matrix occurred in the vicinity of areas with a high fiber concentration. It was also observed that the stress concentrations shifted around the fibers to the location of the interfacial crack tip as a result of the fiber and matrix

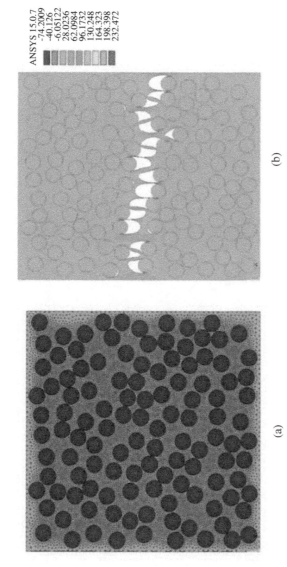

(a)

(b)

Figure 3.10 (a) Finite element model of composite representative unit cell (RUC) and (b) stress contour of RUC at failure.

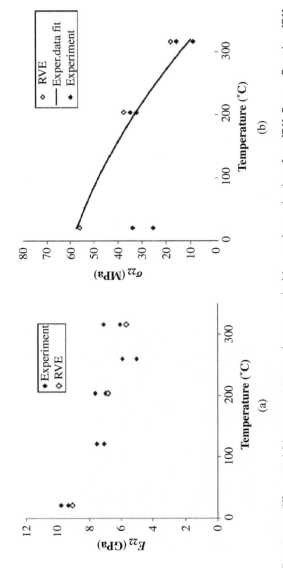

Figure 3.11 (a) Transverse stiffness and (b) transverse strength compared with experimental values from [76]. Source: Based on [76].

becoming completely debonded. Results for transverse stiffness were close to those from experiment [76] for both low and high temperatures, as shown in Figure 3.11a. With respect to changes in temperature, a decrease in stiffness was observed with an increase in temperature, which was expected largely due to the decrease in stiffness of the polyimide matrix. The strength of the composite, shown in Figure 3.11b, was also obtained. Similar to the stiffness, there was a reduction in strength as the temperature increased. This was attributed to the lower maximum normal and shear stress as obtained from the multiscale cohesive zone model. Results agreed with experimental values; however, the lower strength at 21 °C, as measured in [76], can also be attributed to the specimen breaking at the grip as opposed to at the gauge location during the tensile tests. When comparing the strength with the experimental data fit provided in [76], the strength calculated from the RVE agreed well. Overall, the potential of using an atomistic-to-continuum simulation to obtain viable cohesive zone parameters for microstructural models was demonstrated [55].

3.6 Conclusion

In this chapter, an overview is provided on multiscale methods and their applications to lightweight material and structure design. In particular, the BCM multiscale approach is reviewed and examples of its application in lightweight design are presented. In general, multiscale methods are becoming more prevalent in industry as new technologies stemming from nanotech and additive manufacturing open doorways to designing for a desired structural level behavior beginning at the nano and microscales. To further progress the implementation of multiscale simulation in practical applications, future development of the BCM and other multiscale methods should focus on optimizing codes, generalizing the methodology, and combining existing multiscale methods into integrated methodologies. Many exciting developments in multiscale modeling are expected in the coming years, with the continued growth of multiscale modeling playing an important role in the effort to produce and analyze new technological advances.

References

1 Hedayati, R., Hosseini-Toudeshky, H., Sadighi, M. et al. (2018). Multiscale modeling of fatigue crack propagation in additively manufactured porous biomaterials. *Int. J. Fatigue* 113: 416–427.

2 Tomasi, J., Pisani, W.A., Chinkanjanarot, S. et al. (2019). Modeling-driven damage tolerant design of graphene nanoplatelet/carbon fiber/epoxy hybrid composite panels for full-scale aerospace structures. AIAA Scitech 2019 Forum, p. 1273.

3 Budarapu, P.R., Zhuang, X., Rabczuk, T., and Bordas, S.P. (2019). Multiscale modeling of material failure: theory and computational methods. In: *Advances in Crystals and Elastic Metamaterials*, 1e, vol. 52 (ed. M.I. Hussein), 1–103. Elsevier.

4 Zheng, X., Smith, W., Jackson, J. et al. (2016). Multiscale metallic metamaterials. *Nat. Mater.* 15 (10): 1100–1106.

5 Imediegwu, C., Murphy, R., Hewson, R., and Santer, M. (2019). Multiscale structural optimization towards three-dimensional printable structures. *Struct. Multidiscip. Optim.* 60 (2): 513–525.

6 Kochmann, D.M., Hopkins, J.B., and Valdevit, L. (2019). Multiscale modeling and optimization of the mechanics of hierarchical metamaterials. *MRS Bull.* 44 (10): 773–781.

7 Engineering CoICM (2008). *Integrated Computational Materials Engineering: A Transformational Discipline for Improved Competitiveness and National Security*. Washington, DC: National Academies Press.

8 Panchal, J.H., Kalidindi, S.R., and McDowell, D.L. (2013). Key computational modeling issues in integrated computational materials engineering. *Comput. Aided Des.* 45 (1): 4–25.

9 Markl, M. and Körner, C. (2016). Multiscale modeling of powder bed–based additive manufacturing. *Annu. Rev. Mater. Res.* 46: 93–123.

10 Gu, D., Ma, C., Xia, M. et al. (2017). A multiscale understanding of the thermodynamic and kinetic mechanisms of laser additive manufacturing. *Therm. Eng.* 3 (5): 675–684.

11 Iacobellis, V. and Behdinan, K. (2012). Multiscale coupling using a finite element framework at finite temperature. *Int. J. Numer. Methods Eng.* 92 (7): 652–670.

12 Iacobellis, V., Radhi, A., and Behdinan, K. (2018). A bridging cell multiscale modeling of carbon nanotube-reinforced aluminum nanocomposites. *Compos. Struct.* 202: 406–412.

13 Horstemeyer, M.F. (2009). Multiscale modeling: a review. In: *Practical Aspects of Computational Chemistry* (eds. J. Leszczynski and M.K.K. Shukla), 87–135. Springer.

14 Kalamkarov, A.L., Andrianov, I.V., and Danishevs'kyy, V.V. (2009). Asymptotic homogenization of composite materials and structures. *Appl. Mech. Rev.* 62 (3) https://doi.org/10.1115/1.3090830.

15 Penta, R. and Gerisch, A. (2017). The asymptotic homogenization elasticity tensor properties for composites with material discontinuities. *Continuum Mech. Thermodyn.* 29 (1): 187–206.

16 Ramírez-Torres, A., Penta, R., Rodríguez-Ramos, R. et al. (2018). Three scales asymptotic homogenization and its application to layered hierarchical hard tissues. *Int. J. Solids Struct.* 130: 190–198.

17 Fish, J. and Shek, K. (1999). Finite deformation plasticity for composite structures: computational models and adaptive strategies. *Comput. Meth. Appl. Mech. Eng.* 172 (1–4): 145–174.

18 Fish, J., Yu, Q., and Shek, K. (1999). Computational damage mechanics for composite materials based on mathematical homogenization. *Int. J. Numer. Methods Eng.* 45 (11): 1657–1679.

19 Oskay, C. and Fish, J. (2007). Eigendeformation-based reduced order homogenization for failure analysis of heterogeneous materials. *Comput. Meth. Appl. Mech. Eng.* 196 (7): 1216–1243.

20 Kanouté, P., Boso, D., Chaboche, J., and Schrefler, B. (2009). Multiscale methods for composites: a review. *Arch. Comput. Meth. Eng.* 16 (1): 31–75.

21 Hill, R. (1963). Elastic properties of reinforced solids: some theoretical principles. *J. Mech. Phys. Solids* 11 (5): 357–372.

22 Ghosh, S. and Moorthy, S. (1995). Elastic-plastic analysis of arbitrary heterogeneous materials with the Voronoi cell finite element method. *Comput. Meth. Appl. Mech. Eng.* 121 (1–4): 373–409.

23 Pivovarov, D., Zabihyan, R., Mergheim, J. et al. (2019). On periodic boundary conditions and ergodicity in computational homogenization of heterogeneous materials with random microstructure. *Comput. Meth. Appl. Mech. Eng.* 357: 112563.

24 Moulinec, H. and Suquet, P. (1998). A numerical method for computing the overall response of nonlinear composites with complex microstructure. *Comput. Meth. Appl. Mech. Eng.* 157 (1–2): 69–94.

25 Schneider, M., Ospald, F., and Kabel, M. (2016). Computational homogenization of elasticity on a staggered grid. *Int. J. Numer. Methods Eng.* 105 (9): 693–720.

26 Bansal, M., Singh, I., Patil, R. et al. (2019). A simple and robust computational homogenization approach for heterogeneous particulate composites. *Comput. Meth. Appl. Mech. Eng.* 349: 45–90.

27 Geers, M.G., Kouznetsova, V.G., and Brekelmans, W. (2010). Multi-scale computational homogenization: trends and challenges. *J. Comput. Appl. Math.* 234 (7): 2175–2182.

28 Tadmor, E.B., Ortiz, M., and Phillips, R. (1996). Quasicontinuum analysis of defects in solids. *Philos. Mag. A* 73 (6): 1529–1563.

29 Eidel, B. and Stukowski, A. (2009). A variational formulation of the quasicontinuum method based on energy sampling in clusters. *J. Mech. Phys. Solids* 57 (1): 87–108.

30 Knap, J. and Ortiz, M. (2001). An analysis of the quasicontinuum method. *J. Mech. Phys. Solids* 49 (9): 1899–1923.

31 Shenoy, V., Miller, R., Tadmor, E. et al. (1999). An adaptive finite element approach to atomic-scale mechanics – the quasicontinuum method. *J. Mech. Phys. Solids* 47 (3): 611–642.

32 Dupuy, L.M., Tadmor, E.B., Miller, R.E., and Phillips, R. (2005). Finite-temperature quasicontinuum: molecular dynamics without all the atoms. *Phys. Rev. Lett.* 95 (6): 060202.

33 Kim, W.K. and Tadmor, E.B. (2020). Temporal acceleration in coupled continuum-atomistic methods. In: *Handbook of Materials Modeling: Methods: Theory and Modeling* (eds. W. Andreoni and S. Yip), 805–824. Springer.

34 Broughton, J.Q., Abraham, F.F., Bernstein, N., and Kaxiras, E. (1999). Concurrent coupling of length scales: methodology and application. *Phys. Rev. B* 60 (4): 2391.

35 Kohlhoff, S., Gumbsch, P., and Fischmeister, H. (1991). Crack propagation in bcc crystals studied with a combined finite-element and atomistic model. *Philos. Mag. A* 64 (4): 851–878.

36 Shilkrot, L., Miller, R.E., and Curtin, W. (2002). Coupled atomistic and discrete dislocation plasticity. *Phys. Rev. Lett.* 89 (2): 025501.

37 Belytschko, T., Liu, W.K., Moran, B., and Elkhodary, K. (2013). *Nonlinear Finite Elements for Continua and Structures*. Wiley.

38 Gracie, R. and Belytschko, T. (2009). Concurrently coupled atomistic and XFEM models for dislocations and cracks. *Int. J. Numer. Methods Eng.* 78 (3): 354–378.

39 Xiao, S. and Belytschko, T. (2004). A bridging domain method for coupling continua with molecular dynamics. *Comput. Meth. Appl. Mech. Eng.* 193 (17–20): 1645–1669.

40 Xu, M. and Belytschko, T. (2008). Conservation properties of the bridging domain method for coupled molecular/continuum dynamics. *Int. J. Numer. Methods Eng.* 76 (3): 278–294.

41 Karpov, E., Yu, H., Park, H. et al. (2006). Multiscale boundary conditions in crystalline solids: theory and application to nanoindentation. *Int. J. Solids Struct.* 43 (21): 6359–6379.

42 Qian, D., Wagner, G.J., and Liu, W.K. (2004). A multiscale projection method for the analysis of carbon nanotubes. *Comput. Meth. Appl. Mech. Eng.* 193 (17–20): 1603–1632.

43 To, A.C. and Li, S. (2005). Perfectly matched multiscale simulations. *Phys. Rev. B* 72 (3): 035414.

44 Li, X. and Weinan, E. (2005). Multiscale modeling of the dynamics of solids at finite temperature. *J. Mech. Phys. Solids* 53 (7): 1650–1685.

45 Saether, E., Yamakov, V., and Glaessgen, E.H. (2009). An embedded statistical method for coupling molecular dynamics and finite element analyses. *Int. J. Numer. Methods Eng.* 78 (11): 1292–1319.

46 Rudd, R.E. and Broughton, J.Q. (1998). Coarse-grained molecular dynamics and the atomic limit of finite elements. *Phys. Rev. B* 58 (10): R5893.

47 Curtin, W.A. and Miller, R.E. (2003). Atomistic/continuum coupling in computational materials science. *Modell. Simul. Mater. Sci. Eng.* 11 (3): R33.

48 Miller, R.E. and Tadmor, E.B. (2009). A unified framework and performance benchmark of fourteen multiscale atomistic/continuum coupling methods. *Modell. Simul. Mater. Sci. Eng.* 17 (5): 053001.

49 Iacobellis, V. and Behdinan, K. (2013). Comparison of concurrent multiscale methods in the application of fracture in nickel. *J. Appl. Mech.* 80 (5): 051003.

50 Omelyan, I. and Kovalenko, A. (2013). Multiple time step molecular dynamics in the optimized isokinetic ensemble steered with the molecular theory of solvation: accelerating with advanced extrapolation of effective solvation forces. *J Chem. Phys.* 139 (24): 244106.

51 Kim, W. and Tadmor, E.B. (2017). Accelerated quasicontinuum: a practical perspective on hyper-QC with application to nanoindentation. *Philos. Mag.* 97 (26): 2284–2316.

52 Subramaniyan, A.K. and Sun, C. (2008). Engineering molecular mechanics: an efficient static high temperature molecular simulation technique. *Nanotechnology* 19 (28): 285706.

53 LeSar, R., Najafabadi, R., and Srolovitz, D. (1989). Finite-temperature defect properties from free-energy minimization. *Phys. Rev. Lett.* 63 (6): 624.

54 Sutton, A.P. (1989). Temperature-dependent interatomic forces. *Philos. Mag. A* 60 (2): 147–159.

55 Iacobellis, V. (2016). A bridging cell multiscale methodology to model the structural behaviour of polymer matrix composites. PhD thesis. University of Toronto.

56 Iacobellis, V. and Behdinan, K. (2015). Bridging cell multiscale modeling of fatigue crack growth in fcc crystals. *Int. J. Numer. Methods Eng.* 104 (13): 1200–1216.

57 Liu, B., Huang, Y., Jiang, H. et al. (2004). The atomic-scale finite element method. *Comput. Meth. Appl. Mech. Eng.* 193 (17–20): 1849–1864.

58 Radhi, A. and Behdinan, K. (2017). Contemporary time integration model of atomic systems using a dynamic framework of finite element Lagrangian mechanics. *Comput. Struct.* 193: 128–138.

59 Paris, P. and Erdogan, F. (1963). A critical analysis of crack propagation laws. *J. Basic Eng.* 85 (4): 528–533.

60 Liaw, P.K., Lea, T., and Logsdon, W. (1983). Near-threshold fatigue crack growth behavior in metals. *Acta Metall.* 31 (10): 1581–1587.

61 Hicks, M. and King, J. (1983). Temperature effects on fatigue thresholds and structure sensitive crack growth in a nickel-base superalloy. *Int. J. Fatigue* 5 (2): 67–74.

62 Potirniche, G., Daniewicz, S., and Newman, J. Jr., (2004). Simulating small crack growth behaviour using crystal plasticity theory and finite element analysis. *Fatigue Fract. Eng. Mater. Struct.* 27 (1): 59–71.

63 Esawi, A.M. and El Borady, M.A. (2008). Carbon nanotube-reinforced aluminium strips. *Compos. Sci. Technol.* 68 (2): 486–492.

64 Bakshi, S.R., Lahiri, D., and Agarwal, A. (2010). Carbon nanotube reinforced metal matrix composites – a review. *Int. Mater. Rev.* 55 (1): 41–64.

65 Xiao, S. and Hou, W. (2007). Studies of nanotube-based aluminum composites using the bridging domain coupling method. *Int. J. Multiscale Comput. Eng.* 5 (6): 447–459.

66 Xiao, S. and Hou, W. (2006). Studies of size effects on carbon nanotubes' mechanical properties by using different potential functions. *Fuller. Nanotub. Carbon Nanostruct.* 14 (1): 9–16.

67 Zamani, S.M.M. and Behdinan, K. (2018). Multiscale modeling of the mechanical properties of Nextel 720 composite fibers. *Compos. Struct.* 204: 578–586.

68 Zamani, S.M.M. and Behdinan, K. (2019). The effects of microstructural properties and temperature on the mechanical behavior of Nextel 720 composite fibers: a novel multiscale model. *Compos. Part B* 172: 299–308.

69 Zamani, S.M.M., Iacobellis, V., and Behdinan, K. (2016). Multiscale modeling of the nanodefects and temperature effect on the mechanical response of sapphire. *J. Am. Ceram. Soc.* 99 (7): 2458–2466.

70 Iacobellis, V. and Behdinan, K. (2015). Multiscale modeling of polymer/graphite interface. ASME 2015 International Mechanical Engineering Congress and Exposition. American Society of Mechanical Engineers Digital Collection.

71 Romanowicz, M. (2010). Progressive failure analysis of unidirectional fiber-reinforced polymers with inhomogeneous interphase and randomly distributed fibers under transverse tensile loading. *Compos. Part A* 41 (12): 1829–1838.

72 Hadden, C., Jensen, B., Bandyopadhyay, A. et al. (2013). Molecular modeling of EPON-862/graphite composites: interfacial characteristics for multiple crosslink densities. *Compos. Sci. Technol.* 76: 92–99.

73 Velasco-Santos, C., Martinez-Hernandez, A., and Castano, V. (2005). Carbon nanotube-polymer nanocomposites: the role of interfaces. *Compos. Interfaces* 11 (8–9): 567–586.

74 Akutagawa, K., Yamaguchi, K., Yamamoto, A. et al. (2008). Mesoscopic mechanical analysis of filled elastomer with 3D-finite element analysis and transmission electron microtomography. *Rubber Chem. Technol.* 81 (2): 182–189.

75 Yang, S. and Qu, J. (2014). An investigation of the tensile deformation and failure of an epoxy/cu interface using coarse-grained molecular dynamics simulations. *Modell. Simul. Mater. Sci. Eng.* 22 (6): 065011.

76 Odegard, G. and Kumosa, M. (2000). Elastic-plastic and failure properties of a unidirectional carbon/PMR-15 composite at room and elevated temperatures. *Compos. Sci. Technol.* 60 (16): 2979–2988.

4

Characterization of Ultra-High Temperature and Polymorphic Ceramics

Ali Radhi and Kamran Behdinan

Advanced Research Laboratory for Multifunctional Lightweight Structures (ARL-MLS), Department of Mechanical & Industrial Engineering, University of Toronto, Toronto, Canada

4.1 Introduction

Material properties are highly dependent on the hierarchy of length and temporal scales experienced by the structure, where atomic levels can influence characterized structures through certain features pertaining to atomic species, their orientations and crystal types. Analyzing such features with the underlying atomic group symmetry is necessary to characterize hierarchical properties that are influenced from external stimuli, load rates, and sample sizes. In hierarchical multiscale models, one can obtain certain properties at larger scales by analyzing a smaller scale for material properties and/or mechanics and bridge such properties to the higher scale in a sequential fashion [1]. It depends on the nature of the material on how small a level one can go without sacrificing details to understand the macroscopic behavior. For example, nanocomposites reinforced with carbon nanotubes must be analyzed in an atomic scale to develop a hierarchical relationship with respect to the nanotube's orientation, structure and size for dynamic, mechanical and structural composite properties [1]. Stimuli responsive behavior that alters the crystalline structure may require an analysis of crystalline features. Two common methods of characterizing crystalline structures are centrosymmetry parameter (CSP) [2, 3] and common neighbor analysis (CNA) [4, 5]. Despite being well defined for characterization of local deformation and distinctive atomic arrangements through atomic neighbor topologies, such methods are reported to work for simplified structures (such as face centered cube [FCC], body centered cube [BCC] and hexagonal close packed [HCP] structures) that do not distinguish the atomic space group symmetries. Hence, more advanced crystal characterization methods are desirable when considering structures with non-centrosymmetric, multiple atomic species and multiple space group arrangements.

Ultra-high temperature ceramics (UHTCs) are structures with unique space group arrangements that have generated interest in applications related to thermal barriers, reentry vehicles, and the aerospace and nuclear sectors [6, 7]. These structures were reported to be stable for temperatures up to 3000 °C, where composites of such structures have illustrated further demand in fabricating boride, carbide, nitride and some oxide ceramics.

Advanced Multifunctional Lightweight Aerostructures: Design, Development, and Implementation,
First Edition. Kamran Behdinan and Rasool Moradi-Dastjerdi.

Such properties can be accredited to the strong nature of their covalent bonding, resulting in superior strength, hardness and thermal thresholds. As such, they can be employed in a high thermal fatigue environment, being able to survive high levels of loading and radiation. Research centers, such as National Aeronautics and Space Administration (NASA) and the Italian Aerospace Research Center (Centro Italiano Ricerche Aerospaziali, CIRA), have heavily invested in the development of said structures [8]. Their composites would provide higher reliability, service life and performance. Oxidation is a concern of such structures, where constituents can react with ambient oxygen that may have high impact on thermal and structural properties of the composite [9–11]. Hence, chemical reactions with extreme thermal and ambient environment are much more of a concern than change in crystalline structures of said materials due to their high structural stability.

Alternatively, certain ceramics may experience a crystalline structure phase change due to extreme conditions. Such polymorphic ceramics experience structural evolution of new phases of crystalline groups that can be reversible or irreversible, dependent on the state of thermodynamic equilibrium of the phases' transformation. The anatase phase of titania is an example of a phase transformation [12, 13], where anatase transforms irreversibly to the rutile phase at extreme thermal conditions. The transformation is of considerable interest for applications related to high temperature solar cells, gas sensors, and porous separation membranes [14–16]. Other factors that facilitate the transformation are particle shape/sizes, surface area, heating rates, impurities, and atmosphere. Anatase and rutile experience high photovoltaic performances, structural strengths and stability [13], where anatase is more applied toward solar cells, batteries, and biomedical applications [17–19]. However, anatase is considered a metastable phase of titania, whereas rutile is much more stable and has been widely used for tribological and self-cleaning coating applications [20–22]. The nature of the transformation mechanism starts at the nanoscale, where it involves nucleation and growth processes while having two distinct space symmetry groups even though both anatase and rutile are tetragonal. Hence, a nanoscale characterization approach of such structures would be essential in understanding the extent of the transformation when utilizing a multiscale approach to obtain primary mechanics/deformation properties.

In this chapter, several approaches for characterizing UHTCs and polymorphic structures in terms of crystalline, mechanical and chemical stability to external stimuli are outlined. The work shows advanced methods in crystalline atomic feature detection regardless of atomic species, space symmetry groups, and crystal groups. Also, a chemical kinetic model is reviewed for oxidation in UHTC composites. UHTC structures are a class of material that experience superior properties such that their composites have been widely used in aerospace structures for thermal protection and extreme load resistance [6, 7]. The methods are incorporated in a nanoscale characterization, within a multiscale framework, to understand the effect of transformation reactions on the mechanical performance of resultant structures. The next section outlines the recent work on characterizing UHTCs and their composites. Afterwards, a unique crystalline, multiscale approach is highlighted for characterizing polymorphic structures. Finally, the chapter concludes with comments regarding future work and recommendations for novel research in UHTCs, polymorphic structures and their respective ceramics.

4.2 Crystalline Characterization of UHTCs

As mentioned before, UHTCs are a class of ceramics that can exceed high levels of thermal loading without experiencing a large loss in mechanical performance. Due to the nature of their covalent bonding, applications are allowed with extreme heat and mechanical loading such as hypersonic vehicles or thermal protection systems [23–25]. This is a result of their superior hardness, strength, and thermal toughness, which allows extended periods of time to operate under such conditions. UHTCs commonly have a HCP crystal group with P6/mnm space symmetry group belonging to an AlB_2 structure. Boride type UHTCs are a common type of such structures, where the structure has an alternating layering of close packed aluminum (Al) and graphene-like boron (B) layers as shown in Figure 4.1 for zirconium diboride (ZrB_2). Since this is considered a binary atomic configuration (system with two atomic species), setting up a crystalline characterization tool would prove a difficult task for common methods such as CSP and CNA [4, 5]. Recently, the common neighborhood parameter (CNP) method, proposed by Tsuzuki et al. [26], combined the advantages of both CSP and CNA in order to characterize silicon carbide (SiC) as a binary atomic structure while having a proper centrosymmetric symmetry with clearly distinguished crystal phases. A second neighboring formalism was utilized to enhance the characterization of SiC atomic structure in molecular dynamics (MD) simulations [27, 28], but only same species interactions were assigned as optimal neighboring as some species do not have a fist common neighbor. Cross-species formations were considered less than optimal, where a more recent modification to the CNP was introduced to account for both cross and same species interactions. First, we show the CNP expression as follows [26]:

$$Q_i = \frac{1}{n_i} \sum_{j=1}^{n_i} \left| \sum_{k=1}^{n_{ij}} \left(\mathbf{r}_{ik} + \mathbf{r}_{jk} \right) \right|^2 , \tag{4.1}$$

where n_i is the number of neighbors j of atom i. n_{ij} represents the number of common neighbors k between atoms i and j. $\mathbf{r}$ is a distance vector between any two atoms. This formulation is set to overcome the limitation of CSP in terms of non-centrosymmetric structures and CNA for individual atomic descriptions. The atomic summation between one set of neighbors can be seen in Figure 4.2 for an FCC structure. Figure 4.2a represents the CNP formulation, where it can be seen that a single pair of atoms would have a sum of zero from the defined vector sum. This trend may cause loss of information regarding centrosymmetry when characterizing atoms at the boundary edges or close to corners. In order to enhance the characterization summation, a more recent work has given dominance to the common neighbors over the current neighbor (as seen in Figure 4.2b). This work was recently introduced as the predominant common neighborhood parameter (PCNP) [29]. It can be seen that a vector sum would not be zero anymore when having this "predominant" formulation with the central atom i and common neighbors k being anchors to the vector summation toward neighbor atom j. This can help in observing the common atoms' deformation effect

Figure 4.1 Side and top view of ZrB_2 atoms. B atoms are shown with a chemical bonding in a graphene-like layer.

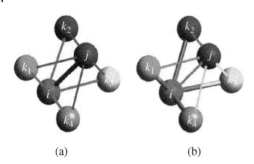

(a)   (b)

Figure 4.2 CNP (a) and PCNP (b) description between two neighbor atoms and their common neighbors through respective vector notations. Source: Reproduced with permission [29]. Copyright 2017, Elsevier.

on the central atom i through the neighbor atom j, which would provide a better distinction to features than the mentioned CNP.

The PCNP formulations are represented by two parameters as follows [29]:

$$P_i = \frac{1}{n_i} \sum_{j=1}^{n_i} \left| \sum_{k=1}^{n_{ij}} \left(\mathbf{r}_{ij} + \mathbf{r}_{kj} \right) \right|^2$$

$$N_i = \frac{1}{n_i} \left| \sum_{j=1}^{n_i} \sum_{k=1}^{n_{ij}} \left(\mathbf{r}_{ij} + \mathbf{r}_{kj} \right) \right|^2 \tag{4.2}$$

where the first parameter P_i is designed to characterize local regions of crystalline structures with a uniform atomic configuration state that helps in visualizing atomic features such as perfect crystalline groups, surface atoms, dislocations, phase changing regions, etc. The second parameter N_i lumps the vector sum of all vectors with neighbors and common neighbors such that it produces a better characterization performance of regions with high local deviations from centrosymmetry such as crack tips/surface, interstitial defects, etc.

Going back to the ZrB_2 structure in Figure 4.1, it can be seen that the graphene-like layers of B atoms do not contain any common first neighbors when considering same species interactions. Moreover, B atoms have 6 Zr and 3 B first neighbor atoms with no opposing atomic configuration, while Zr has 8 Zr and 12 B first neighbor atoms with a centrosymmetric layout. This is common with such UHTC boride structures with space group symmetry of P6/mnm. In order to overcome such an issue, a second neighbor tracking scheme was adopted [28, 29], where the previous formulation can be changed simply by allowing the index j to loop over $n_i^{(l)}$. When l is set to 1, it represents first common neighboring while $l= 2$ is the second common neighboring framework. As mentioned before, the work of Tsuzuki et al. [27] has utilized second common neighboring only for same species interactions. This would not bode well for structures like ZrB_2 when a B layer happens to be on the surface. This would not distinguish surface layers from other layers in the bulk of the atomic domain since the atoms would have a full set of atomic neighbors with similar atomic species. In PCNP, index k is set to cover common second neighbors while keeping j as a set of atom i's first neighbors [29]. This would eliminate the need for an opposing pair of atoms to describe the local state of atomic configuration with only the first neighbor list of any atom i.

With this, an MD study of ZrB_2 with a Tersoff interatomic potential [30] was set up with 5556 Zr and 11 146 B atoms [29]. The simulation was set with a high strain rate of 2.5×10^8 s^{-1} every 10 000 time steps for a total of 220 000 steps with 1 fs time step size. An NPT ensemble was applied initially to the structure for relaxation purposes for 200 000 time

Table 4.1 PCNP value ranges for extracted features in ZrB$_2$ MD simulations.

Feature type	P_i	N_i
HCP (unreformed)	0	0
Dislocation	0.3–0.4	UI
Corner	0.7–0.9	0.25–0.35
Lateral surface	0.1–0.2	0.1–0.2
Crack tip	0.2–0.5	0.7–0.9

UI, Unidentifiable.

steps, followed by NVT ensemble during the loading phase at the <0 0 0 1> direction. The domain size is 104 × 16.5 × 103 Å and the temperature is held at 300 K with the ensembles using a Nosé–Hoover thermostat [31]. Enough time was allowed to observe fracture propagation through the domain, where Figure 4.3 shows the PCNP results after 25 000 time steps in the loading phase of the simulations, where the results are normalized. The first thing to notice is that the results show geometric features regardless of atomic species. P_i results show a clear distinction of the dislocations regions traveling on the [0 0 0 1] plane. The N_i parameter was found to clearly visualize the crack tip with a high contrast from the surrounding dislocation region with no reliance on the type of atoms in this binary atomic system. From such simulations, one can extract the values of the two parameters that correspond to a distinguishable atomic feature observed in the system. Table 4.1 shows the normalized PCNP parameter values and their corresponding atomic features extracted from Figure 4.3.

4.3 Chemical Characterization of a UHTC Composite

UHTCs have been known to experience superior properties in terms of extreme thermal and mechanical loadings with highly reliable/stable structural performances in aerospace components [6, 7]. What is more valuable is that the covalent bonding of said structures provides high chemical stability toward such conditions. This means that the UHTC does not have a crystalline phase change due to high external loading and/or ambient conditions (HCP or P6/mnm do not change). The structures have a melting point of well above 3000 °C [6], where composites of such ceramics experience an improved reliability and a longer service life. When melting or transition occurs at such high temperature levels, there are more critical issues for the composite to face well before such temperature levels are reached. One major issue is oxidation, which is connected to the nature of the extreme conditions typically found in hypersonic and space exploration vehicles. One can have both active and passive oxidation. While active oxidation may cause heavy damage to the structure, passive oxidation may be repurposed as a self-healing agent of such structures [32]. This makes use of the fact that most of the thermal loading experienced for a space vehicle is when existing and entering the atmosphere. For a UHTC composite, the passive oxidation levels interact

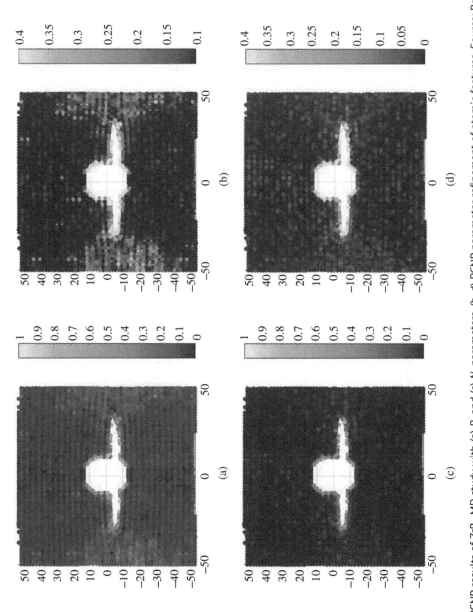

Figure 4.3 PCNP results of ZrB_2 MD study with (a) P_i and (c) N_i parameters. (b, d) PCNP parameter refinement of atomic features. Source: Reproduced with permission [29]. Copyright 2017, Elsevier.

with the additives (commonly SiC) well before the UHTC matrix due to their superior thermal stability. The self-healing feature can be observed by an oxidative chemical reaction: a ZrB$_2$-SiC ceramic was reported to have a recovery in flexural strength after heating to 800 °C [33]. In order to characterize such chemical reactions, one can utilize chemical kinetics laws to describe the rates of oxidation and structure generation and their impact on a constitutive relation for the composite. As mentioned earlier, the additive (in this case SiC) experiences a passive oxidation chemical reaction as follows [34]:

$$2SiC\ (s) + O_2\ (g) \rightarrow 2SiO_2\ (s) + 2CO\ (g) \tag{4.3}$$

The reaction time rate (and consequently the self-healing rate) t_H can be describe in a chemical kinetics law as suggested by Osada et al. [34]. A self-healing frequency rate f_H can thus be outlined as follows [32]:

$$f_H = t_H^{-1}, \text{where } t_H = \frac{1}{k}\left(\frac{P_{O_2}^0}{P_{O_2}}\right)^n = \frac{1}{ka_{O_2}^n} \text{and } k = A\exp\left(-\frac{E_a}{RT}\right) \tag{4.4}$$

The k rate expression follows the Arrhenius law [35] that contains the reaction's activation energy E_a and temperature T. A and R represent the frequency factor and Boltzmann constant, respectively. The partial pressure of O$_2$ (P_{O_2}) is an essential factor the influences the oxidation reaction, where the ratio of P_{O_2} and atmospheric pressure $P_{O_2}^0$ (a_{O_2}) is inversely proportional to t_H in the order of O$_2$ reaction order n.

The above reaction rate was incorporated through a quasi-static damage model to represent the degree of strength recovery. A damage state variable D is defined to describe the damage of each element (0 undamaged; 1 fully damaged) in a constitutive relation as shown [36]:

$$\sigma_{ij} = (1 - D)\,C_{ijkl}\varepsilon_{kl} \tag{4.5}$$

The stress σ_{ij} and strain ε_{kl} tensors are related through the damage state variable D and the material tensor C_{ijkl} obtained from a predefined representative volume element (RVE) of a ZrB$_2$-30% SiC composite using a cohesive zone model (CZM) [32]. The damage state parameter is expressed as follows [36]:

$$D\,(\alpha) = 1 - \frac{\alpha_0}{\alpha}\exp\left(-\frac{E\alpha_0 h_e}{G_f}\left(\alpha - \alpha_0\right)\right) \tag{4.6}$$

The D factor is set as a function of maximum deformation history state variable α, where it is set to have a damage description when exceeding the damage initiation strain α_0. The RVE is also utilized to obtain the fracture energy G_f and Young's modulus E for an element with a characteristic length of h_e. For the purposes of including the chemical kinetic relations, state variable α is divided into two components: a loading strain (α_L) and a self-healing strain ($-\alpha_H$). The negative sign in the self-healing component acts to thwart the damage effect while the loading part is set to the modified von Mises model strain [37] during loading and to itself during unloading (to represent the history of deformation beyond the von Mises model strain). Hence, one must take care if the structure is to be reloaded since the

Table 4.2 Values for material properties and self-healing parameters for ZrB$_2$-30% SiC beam simulation.

Material properties			
E (GPa)	500	v	0.2
G_f (J/m^2)	58	Y	10
α_0	0.0022	α_s	0.0022
Self-healing properties [34]			
A (s^{-1})	1.04×10^{10}	E_a (J/mol)	387 000
R (J/mol/K)	8.314	n	0.557

reloading stage may contain alternative material propagation models. The self-healing frequency rate f_H influences the time rate of maximum deformation history as follows [38]:

$$\dot{\alpha}_H = \varphi_1 H f_H \left(\alpha - \alpha_0 \right) \quad \text{if } \alpha_L > \alpha_0$$
$$\dot{\alpha}_n = \varphi_2 H f_H \left(\alpha_n - \alpha_s \right) \quad \text{if } \alpha_L > \alpha_0$$
$$\text{where } H = \frac{\delta_{\max}}{\left(\alpha_L - \alpha_0 \right) h_e} \quad (4.7)$$

The time rate of the self-healing component $\dot{\alpha}_H$ expression is set to have α_H increase monotonically as α approaches α_0. This is also true for the reloading stage for $\dot{\alpha}_n$ with α_n being the largest α_L beyond α_0 and a new damage initiation strain α_s. The factors φ_1 and φ_2 control the reaction rates for a crack opening displacement H relative to maximum crack opening displacement $\delta_{\max}$.

Table 4.2 shows the utilized material properties in the self-healing constitutive model obtained from the composite RVE. The current study shows a three-point bending simulation with 0.5 mm/min load rate and $36 \times 4 \times 3$ mm beam dimensions as shown in Figure 4.4. $\varphi_{1,2}$ are set to 5 while fracture energy G_f was taken from [39]. The present study utilized 8-node hexahedral elements, where some elements in the shaded domain are set with an initial D value of 0.3 to represent an initial notch for the three-point-bending study with 912 elements and 1365 nodes. The stages of the simulations are loading/unloading, rest and then reloading. The temperature was held at 1073 K for 3600 s at the rest stage with an O$_2$ partial pressure ratio of 0.21. The constitutive relations have been coded in FORTRAN and visualization is processed using ABAQUS.

Figure 4.5 shows the damage state variable D distribution results at the start and end of each stage. Figure 4.5a is the initial setup from Figure 4.4 just after the initiation of the loading stage for 10 s, and Figure 4.5b represents the final state of the three-point bending test at the loading stage with a visible propagation of the notch crack. Figure 4.5c shows the

Figure 4.4 Three-point bending test configuration with applied loads/boundary conditions.

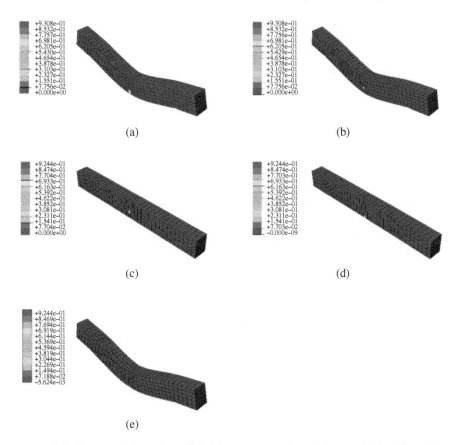

Figure 4.5 Damage state variable D distribution results for the three-point bending study at (a) initial, (b) end of loading, (c) start of rest stage, (d) end of rest stage and (e) end of reloading states. Source: Reproduced with permission [32]. Copyright 2019, Elsevier.

beam after unloading it, which is the same as the start of the rest stage such that the beam is allowed to interact with the ambient oxygen for 3600 s and a self-healing behavior becomes quite visible as the crack heals, illustrated in Figure 4.5d. Figure 4.5e shows the results after 10 s of reloading the beam after rest, where there is minimal propagation of crack as a result of the self-healing behavior compared with Figure 4.5b. The flexural strengths are compared during loading and reloading stages as presented in Table 4.3. It can be clearly seen that there is a recovery in flexural strength with great accuracy compared with the experimental data, attesting to the capability of the presented kinetic/mechanical constitutive relation.

4.4 Polymeric Ceramic Crystalline Characterization

At this point, the presented research on UHTCs attests to the stability of the structure and their reliability against high levels of thermal or mechanical loading. From an atomistic and a continuum perspective, UHTCs and their composites represent a case where certain environmental stimuli may cause such structures to experience different trends (such as

Table 4.3 Calculated flexural results from the three-point-bending test of the composite study (in MPa).

Stage	Simulation values	Experimental values [33]
Loading	675	680
Reloading	746	750

Source: Reproduced with permission [32]. Copyright 2019, Elsevier.

oxidation) well before any extensive crystalline changes occurs (such as phase transition and/or melting). However, this is not the case for all ceramics as some ceramics may experience a "polymorphic" behavior. More specifically, a phase change or transformation of some of the crystalline phases may arise due to an external/environmental load. Such situations may require a more in-depth analysis of the crystalline behavior, degree of change, the mechanism of transformation and their impact on the mechanical performance of the macroscopic structure.

Such transformations are most relevant at surface regions, where cases such as energetic or solar cells are most relevant when having a consistent crystalline structure that does not compromise its performance. Titanium dioxide (TiO_2) is one such example, having multiple polymorphs each with unique biocompatibility and photovoltaic affinity [13]. Rutile is one polymorph of titania with a tetragonal crystal group and a space symmetry group of $P4_2/mnm$ with the atomic layout depicted in Figure 4.6. Ti atoms in the structure have 6 O atoms as first nearest neighbors. O atoms have 3 Ti first nearest neighbors, where both species do not have common first neighbors. As such, conventional methods like CSP and CNA fall short in describing such structures. Moreover, all first neighbors are of different species from the current central atom. If second nearest neighboring was used (much like what was done in Section 4.2 with PCNP), all second neighbors become atoms of the same species. Such non-monoatomic systems can be defined for both same and cross species interactions when taking the second neighbors as the first neighbor of the first neighbor, minus the central atom. Cross species interactions can thus be included indirectly.

Most of the conventional atomic characterization methods utilize a low value range to describe regions of no or minimal deformation/crystalline phases with high regions describing defects, cracks, surface atoms, etc. [40]. More recently, a novel method has changed such convention by awarding high locale of atoms with high values and low distribution of atoms with lower values in the cumulative common neighborhood parameter (CCNP) [41]. As the name suggests, the parameter "cumulatively" lumps displacement vector norms to itself in additive fashion from each neighbor and their common neighbors. The set-up updates both CNP and PCNP schemes to achieve such lumping. The CCNP version of CNP is introduced as [41]:

$$AQ_i = \frac{1}{n_i}\sum_{j=1}^{n_i}\left(\sum_{k=1}^{n_{ij}}\left|\mathbf{r}_{ik}+\mathbf{r}_{jk}\right|\right)^2, BQ_i = \frac{1}{n_i}\left(\sum_{j=1}^{n_i}\sum_{k=1}^{n_{ij}}\left|\mathbf{r}_{ik}+\mathbf{r}_{jk}\right|\right)^2 \tag{4.8}$$

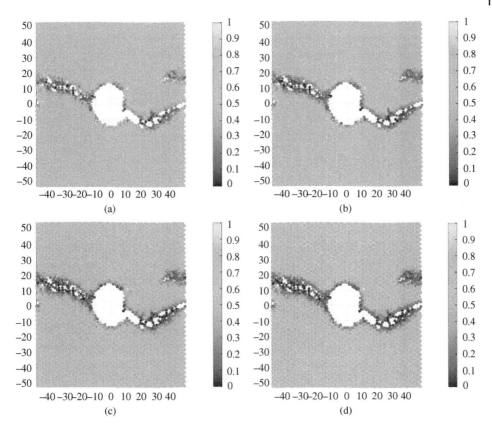

Figure 4.6 CCNP results of rutile MD study with factors (a) AQ_i, (b) BQ_i, (c) AP_i, and (d) BP_i.

And the CCNP modification on PCNP produces this set of parameters [41]:

$$AP_i = \frac{1}{n_i} \sum_{j=1}^{n_i} \left| \sum_{k=1}^{n_{ij}} \left(\mathbf{r}_{ij} + \mathbf{r}_{kj} \right) \right|^2 , BP_i = \frac{1}{n_i} \left| \sum_{j=1}^{n_i} \sum_{k=1}^{n_{ij}} \left(\mathbf{r}_{ij} + \mathbf{r}_{kj} \right) \right|^2 \tag{4.9}$$

One can notice that the vector norms are obtained for each pair of neighbors, common neighbors and central atom to be lumped for each current atom i factor. This provides a steadier characterization of surfaces from its surrounding bulk atoms. The difference between the A factors from the B ones is the difference in amplification or demagnification for a better way to distinguish certain features.

Much like PCNP, CCNP can utilize nearest second neighbors by allowing j to go over $n_i^{(l)}$ second neighbors when $l= 2$ [41]. This is essential when characterizing structures such as the mentioned rutile polymorph. An MD simulation is introduced with 5984 Ti and 11 968 O atoms using a Matsui–Akaogi interatomic potential [42]. The study is set to use CCNP to characterize crack propagation by introducing a 10 Å hole at the center along the <0 1 0> direction of a simulation domain of $101 \times 18.5 \times 101$ Å. The system is initialized with a NPT ensemble for 200 000 time steps, followed by 220 000 load steps with a strain rate of 2.5×10^8 s^{-1} applied every 100 steps with a Nosé–Hoover thermostat [31]. The temperature is held at 300 K; the results in Figure 4.6 are obtained after 53 000 load steps. The

Table 4.4 CCNP value ranges of rutile MD feature study.

Characterization parameter	Tetragonal (undeformed)	Surface	Corner	Crack tip	Dislocation
P_i	0.6–0.8	0.5–0.6	0–0.1	0.2–0.4	0.4–0.5
N_i	0–0.1	0.14–0.2	0.1–0.12	0.4–0.5	UI
AQ_i	> 0.75	0.2–0.4	0–0.25	0.25–0.4	0.4–0.6
BQ_i	> 0.8	0.2–0.3	0–0.1	0.3–0.45	0.4–0.6
AP_i	0.5–0.6	0.25–0.35	0–0.1	0.35–0.45	0.4–0.55
BP_i	0.5–0.65	0.2–0.35	0–0.1	0.25–0.4	0.4–0.55

UI, Unidentifiable.

values in the figure are normalized, where a person can clearly see that bulk features are awarded with parameter values in the high range of the normalized factors and features such as surface, crack tips/corners or suspended chain atoms have values in the lower range (evidence of the cumulative framework of the presented factors). This unique framework allowed the distinction of crack surfaces from the dislocations surrounding the surface with other features such as edge atoms, from deformed or undeformed surfaces. The CNP-based parameters in the figure show a clear distinction of the mentioned features while PCNP modifications illustrate a definite contrast between surface and bulk atoms. However, all sets of parameters share the outcome of bulk atoms having large parametric values, unlike other characterization factors, as a result of the cumulative framework. Table 4.4 presents the obtained features and their corresponding CCNP value ranges.

4.5 Multiscale Characterization of the Anatase–Rutile Transformation

Polymorphic structures can experience a change in their crystalline configurations due to external stimuli without necessarily being of chemical nature. One such transformation is the transformation of anatase to rutile polymorphs of titania [13]. Each polymorph has photovoltaic, strength and stability performances that are unique to the corresponding polymorph. What is interesting about the two polymorphs is that they are both charac- terized with a tetragonal crystal group. However, rutile and anatase have space symmetry groups of P4$_2$/mnm and I4/amd. Up to this point, all characterization parametrization reported in the literature do not distinguish space symmetry groups within the sane crystal group. If one is to be able to characterize the transformation from an atomistic perspective, a novel parameterization must be able to handle the difference between the two space groups. This is the purpose of the CCNP method, which is to characterize rutile from anatase as the phase transformation occurs at critical surface planes (where CCNP excels in terms of feature extraction).

The major factor for the phase transformation is temperature [13]. The metastable anatase transforms irreversibly to stable rutile in nucleation and growth processes. This

Figure 4.7 MD domain setup for anatase to rutile transformation study. The black vertical lines correspond to periodic boundary conditions.

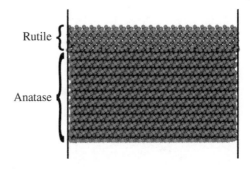

Rutile {

Anatase {

transformation is of critical importance for titania for applications involving gas sensors [43], porous separation membranes [44], and fuel cells [45]. Due to the scale of transformation, one must have an atomistic-level description of the structure and the polymorph's corresponding atomic configurations and lattice parameters. We set up an MD study with 1 fs and a Matsui–Akaogi interatomic potential [42]. The domain is set with both anatase and rutile sharing a grain boundary as seen in Figure 4.7, generated through their respective atomic space symmetry groups. 52 041 O and 14 076 Ti atoms are created and a cut-off radius of 7 Å is used for the potential. The crystalline planes of the grain boundary are [1 0 0] and [1 1 2] plane for rutile and anatase, respectively. They are made to coincide because the respective planes are the critical planes of transformation [13]. The thickness of the rutile layer is 6.88 Å, while anatase spans the rest of the layer (28.15 Å in height). A conjugate-gradient minimization is used before initializing the system at 300 K with a Nosé–Hoover thermostat [46] during 100 000 time steps with another 100 000 steps at 300 K in NPT ensembles. Then, the system is heated for another 100 000 steps to 1300 K and kept there for another 100 000 steps before reducing the system back to 300 K within 100 000 steps.

The results in Figure 4.8 and Figure 4.9 show the initial and final structures of the MD simulation, analyzed by CCNP modifications on CNP and PCNP, respectively. Both figures show a visible increase in the rutile layer thickness, evidence of the irreversible transformation at the grain boundary. The grain boundary and anatase surface show sites of nucleation and ensuing transformation between 900 K and 1200 K, which is in line with previously reported observations [47]. The process includes a rearrangement of atomic cells, where a contraction of anatase's c-axis is observed and bounding harmonic surfaces (with $10\,eV/Å^2$ potential constant) assist in the transformation by providing pressure at the top and bottom surfaces. One can compute the product rutile lattice constants and compare with the initialized rutile lattice parameters [48], as illustrated in Table 4.5, for validation purposes.

Unlike UHTCs, the current structure experiences a polymorphic behavior well below the melting point on the atomistic structure level. A multiscale characterization of such structure should include an atomistic-level description of the underlying material domain to represent such changes in crystalline structures and observe the resulting mechanics. Hence, a concurrent multiscale approach is a good candidate to include both localized regions of atomistic levels while having a simplified continuum domain for regions of small mechanical changes [1].

Credited to CCNP, we are able to distinguish the two space symmetry groups of rutile and anatase despite having the same tetragonal crystal group. The two mentioned polymorphs are in the form of small surfaces in the mentioned applications mentioned at the start of

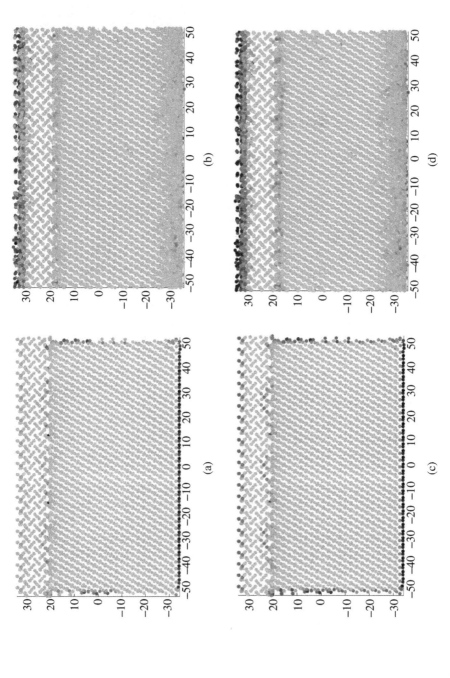

Figure 4.8 (a, b) AQ_i results of titania phase transformation study for initial and final structures, respectively. (c, d) BQ_i results of titania phase transformation study for initial and final structures, respectively.

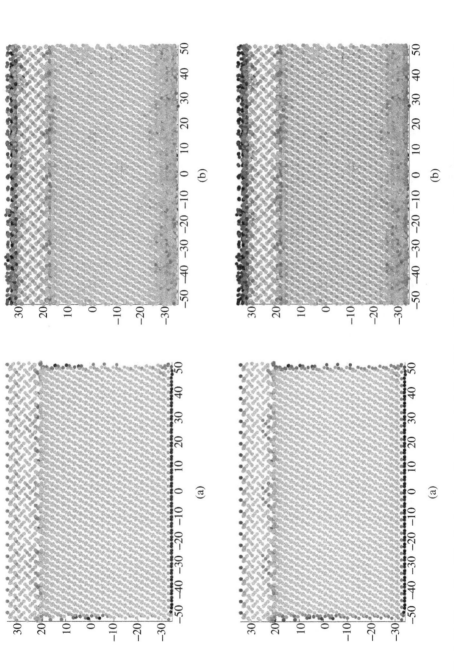

Figure 4.9 (a, b) AP_i results of titania phase transformation study for initial and final structures, respectively. (c, d) BP_i results of titania phase transformation study for initial and final structures, respectively.

Table 4.5 Lattice parameter values of initialized rutile and product rutile from the MD titania phase transformation study.

Lattice parameter	Product rutile	Initialized rutile [48]
a (Å)	4.57	4.587
c (Å)	2.9	2.954

this section. Changes in the polymorphs would adversely change the mechanical properties of the structure, which in turn would affect their performance in their respective applications. Hence, a nanoindentation study is introduced to observe the mechanical properties of the resulting structure on the critical plane of transformation. The initial configuration is shown in Figure 4.10, where an FCC indenter was packed with a lattice spacing of 1 Å with a repulsive Lennard–Jones potential [49] between the indenter and the titania atoms (assigning a potential well depth and zero potential finite distance of 0.1 eV and 1.5 Å, respectively). For the coupling between the atomistic and continuum domain of the two structures, we utilize a dynamic version of the bridging cell method (BCM) [50]. The continuum domain is modeled by conventional finite elements (FEs) [51] in an explicit/implicit numerical integration. The MD implicit solver is adopted from the recent work on the dynamic atomistic finite element (DAFE) framework [52] (run here every 100 steps with 2 fs step size) to have a simplified approach with a conventional FE setup. The coupling between the atomistic and continuum domain is done by a transition/bridging region between the two. The atoms inside this region are mapped to cells/elements that comprise the bridging domain through a mapping function $\overline{\phi}$ by means of inverse isoparametric mapping [53] of the atomic degrees of freedom inside their encompassing cells. The mapping function $\overline{\phi}$ is described as a tensor as follows [50]:

$$\overline{\phi}_{ij} = \begin{cases} \phi_{ij} & i,j \leq n_b \\ \delta_{ij} & i,j > n_b \end{cases}, \tag{4.10}$$

where n_b is the number of atoms in the bridging domain and δ_{ij} is the Kronecker delta function. The above expression is used to obtain the bridging expression between the two domains for mapped mass tensor $M^{(m)}$, stiffness tensor $K^{(m)}$ and load vector $P^{(m)}$ as shown [50, 52]:

$$M_{ij}^{(m)} = \left(1 - w_e\right) m_j \overline{\phi}_{ji} \delta_{jk} \overline{\phi}_{kj} + w_e \delta_{ij} \int_{V_e} \rho N_I N_J dV_e$$

$$K_{ij}^{(m)} = \left(1 - w_e\right) \frac{\partial^2 U}{\partial x_i \partial x_j} + w_e \frac{\partial}{\partial u_{jJ}} \left[\int_{V_e} \frac{\partial \overline{\phi}_{iI}}{\partial x_j} \sigma_{ji} dV_e \right]$$

$$P_i^{(m)} = F_i - \left[\left(1 - w_e\right) \frac{\partial U}{\partial x_i} + w_e \int_{V_e} \frac{\partial \overline{\phi}_{iI}}{\partial x_j} \sigma_{ji} dV_e \right] \tag{4.11}$$

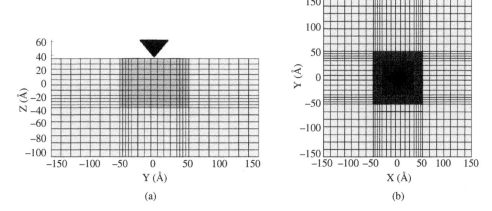

Figure 4.10 Titania multiscale domain discretization for nanoindentation study showing (a) side and (b) top view.

The above set of equations is obtained by applying the principal of virtual work, where a weighting w_e is included in the cells' quadrature points to counteract out-of-balance, non-local energies from atoms inside each cell (1 fully continuum; 0 fully atomistic). As such, the first part of the summation in the previous equation set shows the atomistic contribution with an interatomic potential U. The second part of the summation attributes the continuum component described by a FE framework with shape function N, mass density ρ, displacement vector u, and position vector x. For each cell e with a volume of V_e, a deformation gradient $\psi(\chi_e)$ is obtained while computing the nonlocal contribution (n) of the cell in order to compute the weighting w_e as follows [50]:

$$w_e = 1 - \frac{1/2 u_e^{(n)^T} K_e^{(n)} u_e^{(n)}}{\psi(\chi_e) V_e} \tag{4.12}$$

Applying this weighting would avoid the necessity of refining the transition/bridging cell mesh down to the atomic level, previously adopted by similar methods in the literature [1, 54]. This framework now would allow us to combine the best of continuum and atomistic approaches to characterize the presented nanoindentation study by including appropriate thermostats, simplified FE boundary loads, and high deformation/non-linear solvers.

Next, the nanoindentation simulation is conducted with two indenter speeds: 10 m/s (indenter speed 1) and 20 m/s (indenter speed 2). A value of 3 Å is used as the cut-off radius between the indenter and substrate, where the indenter apex is made to reach 10 Å in the substrate to be held 1 ps before reversing the direction of indentation to relax the structure. The indenter's shape is a four-sided pyramid with 45° as a defining angle [55]. All sides have fixed boundary conditions except for the top side of indentation. The domain contains 12 663 elements with 2132 bridging cells, comprising 16 848 degrees of freedom. In the bridging domain 2820 degrees of freedom are found, where 45 685 atoms are located within the bridging domain such that 45 685 degrees of freedom are reduced to 2820 degrees of freedom due to the multiscale mapping which greatly reduces the computational cost without sacrificing atomic-level descriptions. Figure 4.11 shows the load–depth relation of the current simulation for the two mentioned indentation speeds. The load is obtained by

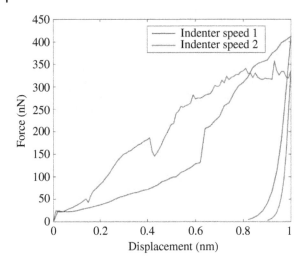

Figure 4.11 Load–depth curve of the titania nanoindentation simulation.

computing all forces experienced by the indenter and the depth is obtained by the rigid indenter's apex position. We compute a property of the nanoindentation, which is called the reduced elastic modulus E_r, as follows [56]:

$$E_r = \frac{\sqrt{\pi}}{2} \frac{S}{\sqrt{A_p}} \tag{4.13}$$

A_p and S refer to the projected area of the indenter and the slope of the unloading part of the load–depth curve at the maximum depth, respectively. We follow the work of Oliver and Pharr [56, 57] to analyze the unload curve and obtain S. The elastic modulus of the substrate can thus be obtained from the following relation [56]:

$$\frac{1}{E_r} = \frac{1 - v_i^2}{E_i} + \frac{1 - v^2}{E} \tag{4.14}$$

E_i is the indenter's elastic modulus with a Poisson's ratio of v_i. The indenter is taken as a rigid one, which leads to having a very large value of E_i compared with the substrate's modulus E. We can thus neglect the indenter's component in the previous relation, and we assign a value of 0.3 for the substrate's Poisson's ratio v. For the sake of comparison, we employ the developed research of Muraishi that characterizes elastic modulus of layered composites by means of nanoindentation [58]. The work is an energy-based/rule of mixture method with each constituent having an energy weighting ratio ϕ instead of a volume fraction. Energy contributions are taken only within the high strain field around the indenter's tip, where Muraishi defines a critical depth R_c relation at which the energy contributions from each layer is computed with R_c defined as [58]:

$$R_c = \sqrt{\frac{\alpha_p}{\pi}} h_c \tag{4.15}$$

α_p and h_c are projected area factor of the indenter and its contact depth, respectively. For a substrate s and n layers below it, the energy weighting ratios have a sum of 1 and the elastic modulus of the layered composite layout E_{com} can be obtained from the following

Table 4.6 Elastic modulus calculations for titania nanoindentation simulation using two indentation speeds compared with the indentation modulus expression developed by Muraishi [58].

Load settings	Simulation modulus E (GPa)	Muraishi modulus E_{com} (GPa)
Indenter speed 1	501.09	486.78
Indenter speed 2	464.72	–

Source: Based on [58].

equation [58]:

$$\frac{1}{E_{com}} = \frac{1}{E_s}\phi_s + \sum_{k=1}^{n}\frac{1}{E_k}\phi_k \tag{4.16}$$

The values of the elastic modulus in the previous expression are obtained from experimental values for anatase and rutile [59–61] oriented at [112] and [100] planes, respectively. Table 4.6 summarizes the elastic modulus results by the multiscale simulation/Oliver and Phar calculations and the empirical/Muraishi formulation of the layered setup. The lower indentation speed has a closer value to the one obtained by the expression developed by Muraishi. However, both speeds have accurate E results with only 5% deviation from E_{com}.

4.6 Conclusion

In this chapter, we introduce some of the recent work involving characterization, analysis and multiscale property calculations of UHTC and polymorphic structures. Both these types of materials have unique thermal, structural and electrical properties due to their crystalline structure, types of bonding, and multiscale nature. UHTCs have especially high tolerance to extreme thermal environments, where they can withstand temperatures up to 3000 °C without change in crystal structure. The novel method of PCNP is utilized to characterize this complex configuration of a crystalline structure for a binary atomic system of UHTC boride material. It is shown to have great feature detection regardless of atomic species and shows structural stability at higher temperatures in an MD study. However, these types of materials are prone to environmental effects well before any transition or melting occurs. One such factor for UHTC composites is the effect of oxidation on the structure, which can happen at temperatures below 1000 °C. A larger multiscale method was utilized with a chemical kinetics constitutive relation described by a damage parameter, an oxidative rate and a hierarchical multiscale formulation. This higher scale was utilized as a result of UHTCs having a stable molecular structure, which eliminates the need of having atomic-level description to characterize the change in material structure. A three-point bending study of a UHTC composite shows the recent work of the chemical kinetic model in describing environmental oxidation of such structures. However, the same could not be said for polymorphic ceramics. A change in crystalline structure is a behavior of polymorphic structures that ensues when exposed to mechanical, thermal, electrical or chemical elements. Anatase is one type of such structure as the metastable ceramic irreversibly transforms to the more stable rutile polymorph of titania. High temperature levels are reported to be the most observable factor of such transformation. Not only a change in atomic configuration is detected, but anatase to rutile transformations have their structures change within the same crystal group having two disparate space symmetry groups at critical surfaces/planes of transformation. The work of CCNP is introduced to highlight its capability of detecting space symmetry groups within the same tetragonal crystal structure of the two titania polymorphs with a high sensitivity toward surface features contrasting between different atomic features. Since the change in the material is observed in the atomic level, a concurrent multiscale approach is utilized with a newly formulated, dynamic-based BCM framework. The framework is utilized in a titania nanoindentation study after characterizing the degree of polymorphic changes along the critical planes by CCNP. These application areas are but a small subset of the potential area of electrical, physical and mechanical characterization of UHTC and polymorphic structures. Some possibilities include thermomechanical characterization, polymorphic inhibitors/facilitators, reactive atomic structures, and fatigue. Each of those possibilities directly play toward the design for manufacturing, environment, compatibility and lightweight structures. Prospective designers are recommended to characterize features related to fracture toughness, chemical/environmental resistivity/compatibility, contact fixtures/joint designs, extreme environment data recording and thermal/mechanical/electrical fatigue. Such characterization is the best approach toward reduced development times, robust testing and improved device properties for UHTCs, polymorphic structures, and their respective composites.

References

1 Tadmor, E.B. and Miller, R.E. (2011). *Modeling Materials: Continuum, Atomistic and Multiscale Techniques*. Cambridge University Press.

2 Kelchner, C.L., Plimpton, S., and Hamilton, J. (1998). Dislocation nucleation and defect structure during surface indentation. *Phys. Rev. B* 58: 11085.

3 Liu, Z. and Zhang, R. (2018). AACSD: an atomistic analyzer for crystal structure and defects. *Comput. Phys. Commun.* 222: 229–239.

4 Faken, D. and Jónsson, H. (1994). Systematic analysis of local atomic structure combined with 3D computer graphics. *Comput. Mater. Sci.* 2: 279–286.

5 Stukowski, A. (2016). Visualization and analysis strategies for atomistic simulations. In: *Multiscale Materials Modeling for Nanomechanics* (eds. C.R. Weinberger and G.J. Tucker), 317–336. Springer.

6 Fahrenholtz, W.G. and Hilmas, G.E. (2017). Ultra-high temperature ceramics: materials for extreme environments. *Scr. Mater.* 129: 94–99.

7 Golla, B.R., Mukhopadhyay, A., Basu, B., and Thimmappa, S.K. (2020). Review on ultra high temperature boride ceramics. *Prog. Mater Sci.* 111: 100651.

8 Ferraiuolo, M., Scigliano, R., Riccio, A. et al. (2019). Thermo-structural design of a ceramic matrix composite wing leading edge for a re-entry vehicle. *Compos. Struct.* 207: 264–272.

9 Marschall, J., Pejakovic, D.A., Fahrenholtz, W.G. et al. (2009). Oxidation of ZrB₂-SiC ultrahigh-temperature ceramic composites in dissociated air. *J. Thermophys. Heat Transfer* 23: 267–278.

10 Carney, C., Cinibulk, M., and Parthasarathy, T. (2017). Processing and evaluation of UHTC loaded composites Ultra-High Temperature Ceramics: Materials for Extreme Environment Applications IV Conference, Windsor.

11 Zhang, L. and Padture, N.P. (2017). Inhomogeneous oxidation of ZrB₂-SiC ultra-high-temperature ceramic particulate composites and its mitigation. *Acta Mater.* 129: 138–148.

12 Gouma, P.I. and Mills, M.J. (2001). Anatase-to-rutile transformation in titania powders. *J. Am. Ceram. Soc.* 84: 619–622.

13 Hanaor, D.A. and Sorrell, C.C. (2011). Review of the anatase to rutile phase transformation. *J. Mater. Sci.* 46: 855–874.

14 Feng, S., Ding, D., Runa, A. et al. (2019). Enhanced photovoltaic property and stability of perovskite solar cells using the interfacial modified layer of anatase TiO₂ nanocuboids. *Vacuum* 166: 255–263.

15 Jia, C., Dong, T., Li, M. et al. (2018). Preparation of anatase/rutile TiO₂/SnO₂ hollow heterostructures for gas sensor. *J. Alloys Compd.* 769: 521–531.

16 Zawrah, M., Khattab, R., Girgis, L. et al. (2017). Synthesis of anatase nano wire and its application as a functional top layer for alumina membrane. *Ceram. Int.* 43: 17104–17110.

17 Khan, J., Rahman, N.U., Khan, W.U. et al. (2019). Multi-dimensional anatase TiO₂ materials: synthesis and their application as efficient charge transporter in perovskite solar cells. *Sol. Energy* 184: 323–330.

18 Guo, Y.-G., Hu, Y.-S., and Maier, J. (2006). Synthesis of hierarchically mesoporous anatase spheres and their application in lithium batteries. *Chem. Commun.*: 2783–2785.

19 Singh, V., Rao, A., Tiwari, A. et al. (2019). Study on the effects of Cl and F doping in TiO₂ powder synthesized by a sol-gel route for biomedical applications. *J. Phys. Chem. Solids* 134: 262–272.

20 Krishna, D.S.R., Brama, Y., and Sun, Y. (2007). Thick rutile layer on titanium for tribological applications. *Tribol. Int.* 40: 329–334.

21 Shamsudin, S., Ahmad, M., Aziz, A. et al. (2017). Hydrophobic rutile phase TiO₂ nanostructure and its properties for self-cleaning application. *AIP Conf. Proc.* 1883: 020030.

22 Haider, A.J., Jameel, Z.N., and Al-Hussaini, I.H.M. (2019). Review on: titanium dioxide applications. *Energy Proc.* 157: 17–29.

23 Squire, T.H. and Marschall, J. (2010). Material property requirements for analysis and design of UHTC components in hypersonic applications. *J. Eur. Ceram. Soc.* 30: 2239–2251.

24 Paul, A., Jayaseelan, D.D., Venugopal, S. et al. (2012). UHTC composites for hypersonic applications. *Am. Ceram. Soc. Bull.* 91: 22–29.

25 Savino, R., Fumo, M.D.S., Paterna, D., and Serpico, M. (2005). Aerothermodynamic study of UHTC-based thermal protection systems. *Aerosp. Sci. Technol.* 9: 151–160.

26 Tsuzuki, H., Branicio, P.S., and Rino, J.P. (2007). Structural characterization of deformed crystals by analysis of common atomic neighborhood. *Comput. Phys. Commun.* 177: 518–523.

27 Tsuzuki, H., Rino, J., and Branicio, P. (2011). Dynamic behaviour of silicon carbide nanowires under high and extreme strain rates: a molecular dynamics study. *J. Phys. D: Appl. Phys.* 44: 055405.

28 Tsuzuki, H (2008). Estudo por dinâmica molecular de deformaçoes mecânicas no cobre e nos semicondutores SiC e InP. PhD thesis. Federal University of São Carlos.

29 Radhi, A. and Behdinan, K. (2017). Identification of crystal structures in atomistic simulation by predominant common neighborhood analysis. *Comput. Mater. Sci.* 126: 182–190.

30 Daw, M.S., Lawson, J.W., and Bauschlicher, C.W. Jr., (2011). Interatomic potentials for zirconium diboride and hafnium diboride. *Comput. Mater. Sci.* 50: 2828–2835.

31 Frenkel, D. and Smit, B. (2001). *Understanding Molecular Simulation: From Algorithms to Applications*. Elsevier.

32 Radhi, A., Iacobellis, V., and Behdinan, K. (2019). A passive oxidation, finite element kinetics model of an Ultra-High Temperature Ceramic composite. *Compos. Part B* 175: 107129.

33 Neuman, E.W., Hilmas, G.E., and Fahrenholtz, W.G. (2013). Mechanical behavior of zirconium diboride–silicon carbide ceramics at elevated temperature in air. *J. Eur. Ceram. Soc.* 33: 2889–2899.

34 Osada, T., Nakao, W., Takahashi, K., and Ando, K. (2009). Kinetics of self-crack-healing of alumina/silicon carbide composite including oxygen partial pressure effect. *J. Am. Ceram. Soc.* 92: 864–869.

35 Levine, R.D. (2009). *Molecular Reaction Dynamics*. Cambridge University Press.

36 Kurumatani, M., Terada, K., Kato, J. et al. (2016). An isotropic damage model based on fracture mechanics for concrete. *Eng. Fract. Mech.* 155: 49–66.

37 De Vree, J., Brekelmans, W., and Van Gils, M. (1995). Comparison of nonlocal approaches in continuum damage mechanics. *Comput. Struct.* 55: 581–588.

38 Ozaki, S., Osada, T., and Nakao, W. (2016). Finite element analysis of the damage and healing behavior of self-healing ceramic materials. *Int. J. Solids Struct.* 100: 307–318.

39 Zimmermann, J.W., Hilmas, G.E., and Fahrenholtz, W.G. (2009). Thermal shock resistance and fracture behavior of ZrB_2–based fibrous monolith ceramics. *J. Am. Ceram. Soc.* 92: 161–166.

40 Stukowski, A. (2012). Structure identification methods for atomistic simulations of crystalline materials. *Modell. Simul. Mater. Sci. Eng.* 20: 045021.

41 Radhi, A., Iacobellis, V., and Behdinan, K. (2018). A cumulative approach to crystalline structure characterization in atomistic simulations. *J. Phys. Chem. C* 122: 13156–13165.

42 Matsui, M. and Akaogi, M. (1991). Molecular dynamics simulation of the structural and physical properties of the four polymorphs of TiO_2. *Mol. Simul.* 6: 239–244.

43 Zakrzewska, K. and Radecka, M. (2017). TiO_2-based nanomaterials for gas sensing—influence of anatase and rutile contributions. *Nanoscale Res. Lett.* 12: 89.

44 Zulhairun, A., Subramaniam, M., Samavati, A. et al. (2017). High-flux polysulfone mixed matrix hollow fiber membrane incorporating mesoporous titania nanotubes for gas separation. *Sep. Purif. Technol.* 180: 13–22.

45 El-Sayed, A., Atef, N., Hegazy, A.H. et al. (2017). Defect states determined the performance of dopant-free anatase nanocrystals in solar fuel cells. *Sol. Energy* 144: 445–452.

46 Martyna, G.J., Tobias, D.J., and Klein, M.L. (1994). Constant pressure molecular dynamics algorithms. *J. Chem. Phys.* 101: 4177–4189.

47 Zhang, H. and Banfield, J.F. (2000). Phase transformation of nanocrystalline anatase-to-rutile via combined interface and surface nucleation. *J. Mater. Res.* 15: 437–448.

48 Muscat, J., Swamy, V., and Harrison, N.M. (2002). First-principles calculations of the phase stability of TiO_2. *Phys. Rev. B* 65: 224112.

49 Tavazza, F., Senftle, T., Zou, C. et al. (2015). Molecular dynamics investigation of the effects of tip–substrate interactions during nanoindentation. *J. Phys. Chem. C* 119: 13580–13589.

50 Iacobellis, V. and Behdinan, K. (2012). Multiscale coupling using a finite element framework at finite temperature. *Int. J. Numer. Methods Eng.* 92: 652–670.

51 Zienkiewicz, O.C., Taylor, R.L., and Zhu, J.Z. (2005). *The Finite Element Method: Its Basis and Fundamentals*. Elsevier.

52 Radhi, A. and Behdinan, K. (2017). Contemporary time integration model of atomic systems using a dynamic framework of finite element Lagrangian mechanics. *Comput. Struct.* 193: 128–138.

53 Murti, V. and Valliappan, S. (1986). Numerical inverse isoparametric mapping in remeshing and nodal quantity contouring. *Comput. Struct.* 22: 1011–1021.

54 Dupuy, L.M., Tadmor, E.B., Miller, R.E., and Phillips, R. (2005). Finite-temperature quasicontinuum: molecular dynamics without all the atoms. *Phys. Rev. Lett.* 95: 060202.

55 Micro Star Technologies (n.d.). Nano Indenters from Micro Star Technologies, Revision 2.3.

56 Oliver, W.C. and Pharr, G.M. (1992). An improved technique for determining hardness and elastic modulus using load and displacement sensing indentation experiments. *J. Mater. Res.* 7: 1564–1583.

57 Yan, W., Pun, C.L., Wu, Z., and Simon, G.P. (2011). Some issues on nanoindentation method to measure the elastic modulus of particles in composites. *Compos. Part B* 42: 2093–2097.

58 Muraishi, S. (2009). Mixture rule for indentation derived Young's modulus in layered composites. *Thin Solid Films* 518: 233–246.

59 Shojaee, E. and Mohammadizadeh, M. (2009). First-principles elastic and thermal properties of TiO_2: a phonon approach. *J. Phys. Condens. Matter* 22: 015401.

60 Iuga, M., Steinle-Neumann, G., and Meinhardt, J. (2007). Ab-initio simulation of elastic constants for some ceramic materials. *Eur. Phys. J. B* 58: 127–133.

61 Momenzadeh, L., Moghtaderi, B., Belova, I.V., and Murch, G.E. (2018). Determination of the lattice thermal conductivity of the TiO_2 polymorphs rutile and anatase by molecular dynamics simulation. *Comput. Condens. Matter* 17: e00342.

Part II

Multifunctional Lightweight Aerostructure Applications

5

Design Optimization of Multifunctional Aerospace Structures

Mohsen Rahmani and Kamran Behdinan

Advanced Research Laboratory for Multifunctional Lightweight Structures (ARL-MLS), Department of Mechanical & Industrial Engineering, University of Toronto, Toronto, Canada

5.1 Introduction

Started in the last century and grown tremendously in the present one, the aerospace industry truly symbolizes the world's current technological competency. It has changed the life of humans in unprecedented ways through enabling fast and safe traveling and transportation across the globe. Yet, the demand for higher performance aircraft has never ceased. With fuel efficiency and lightweighting remaining central to the advancement of the aerospace industry, momentous efforts are focused on novel multifunctional designs, manufacturing technologies, and analysis tools of unparalleled power, all of which serve to accelerate production of novel structures with unusual properties, devise cost-effective and eco-friendly solutions, extend life cycle, and abide by the tightening government legislations [1].

Throughout human history, the standard of living has been impacted by the mastery of materials and structures by humans, to the point that various periods of civilization are branded by their industrious capabilities, such as the Stone Age or Iron Age. The current time may then be well recognized as the Multifunctional Materials Age, as the quest for superior performance has led to mixing materials and integrating components into designs capable of serving more than one purpose. This is largely inspired by nature, as most natural systems are inherently multipurpose and showcase a multilevel design [2].

The aerospace industry has been at the center of the development of architected materials and multifunctional structures to enhance the structural competence of systems, marked by factors such as strength, stiffness, fatigue resistance, energy absorption, and energy dissipation. Advanced manufacturing technologies have also been invented to make such designs possible and cost-effective, the implementation of which has largely dictated achievable speeds, ranges, and payloads of modern aircraft. Yet, the fuel efficiency and environment friendly requirements of the next decade mandated to achieve the ambitious goals put forward by international regulatory bodies calls for re-imagination of conventional system and design in the aerospace sector to move toward lighter and more compact designs. This can largely be addressed through integration of subsystems and multifunctional design practice.

Advanced Multifunctional Lightweight Aerostructures: Design, Development, and Implementation,
First Edition. Kamran Behdinan and Rasool Moradi-Dastjerdi.

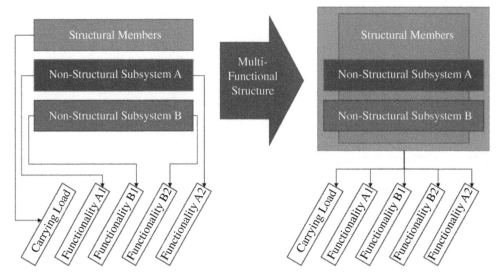

Figure 5.1 Multifunctional structures concept depiction.

This chapter presents an overview of the fundamentals of design and optimization of multifunctional structures and presents example applications in the aerospace engineering realm with an emphasis on recent developments.

5.2 Multifunctional Structures

A multifunctional structure incorporates non-load-bearing subsystems into load-bearing structures with the aim of reducing the total mass and volume of the system [3]. The role of the structural members is then expanded from merely carrying loads, to housing and supporting the operation of the added subsystems. In a multifunctional structure, the functional essence of individual subsystems should be preserved for the resulting system to deliver those functionalities.

Naturally, such efficient designs are sought in the aerospace industry where compactness and lightweighting are major objectives of design optimization efforts. Due to the presence of several subsystems in multifunctional structures, expertise from various engineering disciplines needs to be brought together, often leading to interdisciplinary research campaigns. Examples of multifunctional structures include integration of batteries, thermal control, vibration suppression, health monitoring, noise reduction, the electrical subsystem, and sensors into the load-bearing structures. Figure 5.1 depicts this concept schematically.

To illustrate, one can consider the case of electronic circuits in spacecraft and aircraft, which are conventionally packaged in protective boxes and interconnected through insulated harnesses. The protective boxes are then fixed to structural members in various locations of the vehicle. This is inherently inefficient with respect to overall mass and space required due to the wasted space in between the parts, as well as extra material used for connections and harnesses. To alleviate, the electronics can be miniaturized and implemented in composite sandwich panels used to carry loads in the aircraft [4]. Furthermore, power systems including solar cells and batteries are major contributors to the volume and mass. Embedding thin film batteries [5] in carbon reinforced plastics can offer a compact power solution integrated into the existing structural members. Through multifunctionalization of composite materials to add energy storing functionality, ample structural lightweighting can be realized by reducing masses of non-energy-storing parts, leading to range extension of energy storing units [6].

5.3 Computational Design and Optimization

A central part in development of aerospace structures is computational design and optimization. Design of a new component is an iterative process in nature. However, the current computational tools available to the designers allow for the process to be considerably streamlined and accelerated. Before fabricating prototypes and testing them at component level, the proposed concepts will be computationally modeled and analyzed for various use cases anticipated for the part. Traditionally, various features of the components (such as its topology, shape, material, etc.) will be determined in a largely sequential manner, leading to a final design which meets all the requirements. The challenge of lightweighting and compactness is also addressed in this process. However, often a lighter design might be possible if the search for optimal solution could be elongated or architected materials could be used in the construction. Use of commercial optimization packages can help the designer explore the design space more thoroughly and efficiently. However, the effort required to initiate and set up such procedures can discourage practitioners from following this route. Yet, using techniques such as topology optimization has become the norm in the development of most structures, with the aerospace and automotive sectors pioneering it.

With multifunctional structures, meeting the design requirements can pose a greater challenge as the functionalities of the subsystems should to be preserved to a great extent for the multifunctional design to be competitive. Typically, different functions of the design are conflicting in nature, leading to trade-offs in performance of one aspect or another. Yet, the overall reduction in mass and volume, accompanied by potential cost reduction resulting from the integration of modules and lesser part numbers can render the design sufficiently attractive to be implemented in practice. The challenges of designing novel and efficient multifunctional structures are then manifested in selecting the proper material type (with architected or composite materials being carefully considered), finding an optimized material distribution for the load-bearing components, and integrating the non-load-bearing parts.

In tackling a structural design task, material distribution techniques are indispensable. These include topology, shape, and size optimization, all of which are available to some extent in the commercial software for design optimization. Introduced more than three decades ago [7], topology optimization is a powerful tool to establish the optimal material layout and hence the optimal load path, subject to constraints, applied loads, and boundary conditions. The development of topology optimization has been largely driven by ambitious lightweighting targets in the aerospace and automotive industries. It has grown to become a standard tool employed by engineers [8].

The key advantage of topology optimization over shape and size optimizations is that no structural topology needs to exist a priori. Instead, in this technique, the available design space can be represented using discretized volumes of material commonly referred to as elements, where the elements comprising the optimized layout can be found through solving the underlying mathematical optimization problem [9, 10]. Originally formulated as a binary design problem, where each element can either be full or a void, the computational challenges arising from the large-scale integer programming problem led to alternative approaches. The most widely used density-based topology optimization approach is known as the Solid Isotropic Material with Penalization (SIMP) method [9]. Here, a virtual density value is assigned to each element which characterizes its stiffness and mass contribution. A generic formulation for a topology optimization targeting compliance reads:

$$\text{Minimize } C\left(\rho_i\right) \tag{5.1}$$

$$\text{Subject to } \begin{cases} K_L\left(\rho_i\right) U_L = f \\ \sum_{i=1}^{N} \rho_i v_i \le V_f^* \\ 0 < \rho_{\min} \le \rho_i \le 1 \\ \rho_i = g_j\left(d_i\right) \end{cases} \tag{5.2}$$

where function C represents the compliance of the structure and the aim is to minimize it to achieve the least deformation under the applied external load f. The design variables are ρ_i parameters, which designate the virtual density of the ith element. The ρ_i parameters are unitless, continuous, and real quantities such that $V_f = \sum_{i=1}^{N} \rho_i^k v_i$ [10]. Here V_f is the targeted volume fraction, and v_i is the volume of the ith element in the design space consisting of N elements. K_L represents the linear global stiffness matrix and U_L stands for the nodal displacement vector. A target fraction of V_f^* can be enforced through the second constraint. A minimum density of $\rho_{\min}$ is normally imposed [10] as described by the third constraint, to circumvent singularity in the finite element (FE) problem. Other constraints such as symmetry or manufacturing considerations to ensure a specific pattern can be described by appropriate choices of g_j functions in the last constraint. Figure 5.2 shows a benchmark compliance design task for which topology optimization leads to the truss structure presented.

Elements with virtual density values in proximity of zero can be considered as voids, while those with densities approaching one are construed as solid areas. However, intermediate density values are not obvious to interpret. Therefore, these values are directed to converge toward either the upper or lower bounds to address this challenge in the SIMP method. This is achieved through implementing a penalty parameter P:

$$E = \left(\rho_i\right)^P E_0 \tag{5.3}$$

where E and E_0 represent the penalized and original Young's modulus of the material. It is shown that $P \ge 3$ should hold for this approach to work [10]. Hence, the value of $P = 3$ is frequently used when implementing SIMP [9, 11, 12].

Due to the large number of design variables and the computational cost associated with nonlinear function evaluations and sensitivity calculation, using nonlinear analysis in the topology optimization is impractical [13]. Therefore, in cases where the external loads on

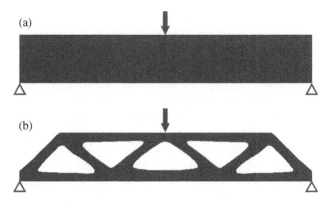

Figure 5.2 The design space (a) and topologically optimized structure (b) of a benchmark compliance design problem.

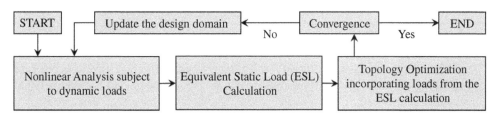

Figure 5.3 Topology optimization procedure incorporating ESL.

the part are time dependent, the standard topology optimization based on linear static analysis can result in a suboptimal layout [11]. To address this challenge, Equivalent Static Load (ESL) methods have been developed [14] and implemented in some of the commercially available software too [15]. The ESL is computed so that it results in the same system response as the actual dynamic loads. Topology optimization with dynamic loads will then be possible by calculating the ESL in each optimization loop, as shown in the flowchart in Figure 5.3. A comprehensive overview of ESL-based topology optimization and several examples of its application to structural design problems of different complexity is provided by Park [14]. Wong et al. [11] performed optimization of the slave links of landing gear using Multi-body Dynamic (MBD) analysis with deformable bodies subject to dynamic loads. Cases of optimization studies involving material, geometry, and contact nonlinearities can be found in the work of Ahmad et al. [13].

Satisfying stress constraints in a topology optimization poses considerable challenge. Such constraints are computationally expensive for gradient-based optimizers, given the number of active constraints are potentially in the order of the number of design variables [16]. Singularity of the global or local stress responses can occur in low-density elements, where excessive distortions are likely due to the lowered local stiffness of the material. Using a special node selection algorithm to calculate the strain–displacement matrices of irregular-shaped elements, Jeong et al. [17] proposed a method for efficient treatment of local stress constraints which does not require access to the internal FE data. Defining a global stress constraint is an alternative approach common in commercial software, where only the maximum stress at each iteration is monitored with the aim of keeping it under a given limit. Functions such as a *p*-norm can then be incorporated as the stress constraint aggregation method. When working with simple structures such as a two-dimensional L-bracket design subject to multiple load cases, the level set approach [18] is reported to be successful in enforcing several stress criteria.

Another shortcoming of the density-based topology optimization method is the formation of checkerboard patterns. These are regions with semi-dense elements which are a challenge to interpret and manufacture. It is suggested that such patterns are in fact numerical noise rather than optimum microstructures [16]. Current commercial structural optimization solvers offer checkerboard suppression algorithms which can combat this effect, while using second-order elements in the FE mesh is also effective in resolving this issue. For a wide class of structural optimization problems [16], reducing the FE mesh size typically leads to the formation of small size members. Hence, the outcome of the topology optimization can depend on the FE mesh used. Imposing a minimum member size constraint is then crucial to ensure simplicity of the design and its affordable manufacturability.

5.4 Applications

Topology optimization is widely used for aerospace structure design, such as in the case of designing an Airbus pylon developed through a bi-level scheme [19]. Multifunctional metamaterials with exotic thermomechanical characteristics have been developed using level-set topology optimization [20]. In order to manufacture multifunctional components in a single operation in an additive manner, topology optimization can be incorporated in the structural design coupled with the system design process [21] or used with nature-inspired models such as a Lindenmayer System [22]. Thermal management is a significant barrier to weight reduction in aircraft which can be alleviated using thermo-electric generators which reduce the amount of heat to be managed by converting it to electrical energy. To this end, topology optimization has been employed in the design of compliant thermoelectric generator legs [23]. Strain-sensing nanocomposite structures are also developed using topology optimization [24].

To illustrate further, next we elaborate on a relevant case study of computational design optimization of a multifunctional aerospace structure. The case presented covers the structural design of a novel shimmy suppression mechanism for aircraft nose landing gears (NLGs). The conceptualization, parametric studies, and optimization of this novel shimmy damper are presented in more detail in previous publications [12, 25–28].

5.4.1 Design Optimization of a Novel NLG Shimmy Damper

In aircraft landing gears, the term shimmy describes self-sustained rotational-lateral vibrations which are observed in some nose and main landing gears during ground operations such as landing [29]. It is crucial to effectively predict and mitigate such oscillations, since they can interfere with the control of the aircraft on the ground, cause discomfort to occupants, and lead to structural failures of the gear due to the cyclic loads [30, 31]. Various factors can influence and invoke shimmy, including insufficient torsional stiffness and damping [32], freeplay in the joints [33], low friction in the shock absorber [34], and low tire pressure [35].

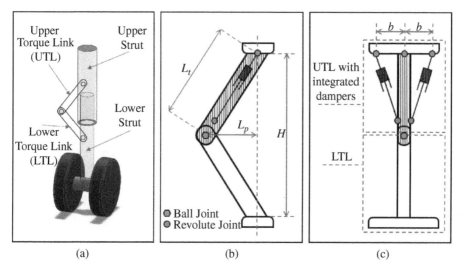

Figure 5.4 Schematics of the NLG structure (a), STLD side view (b), and front view (c).

This challenge is addressed largely by addition of vibration suppression systems known as shimmy dampers, which could be active [36], semi-active [37], or passive [38]. Installed externally on the struts or internally in the steering collar, passive shimmy dampers are the prevalent solution due to their simplicity and not requiring a power source. A novel shimmy damper has been conceptualized [25, 28] and designed [12] which integrates the shimmy suppression device into the torque link system, hence forming a multifunctional structure. The conventional torque link system consists of two structural members (upper and lower) which essentially transfer the torque from the wheels and lower strut to the upper strut of the NLG. In the proposed multifunctional system, designated as the Symmetric Torque Link Damper (STLD), the torque transfer functionality is preserved while vibration energy dissipation functionality is also added by incorporating the damping elements. Figure 5.4 depicts a simplified NLG structure and the STLD concept schematically. The working principle of STLD is dissipation of oscillatory energy in the damping elements mounted symmetrically on the two sides of the upper torque link (UTL). Upon rotation of the UTL beam as a result of rotational motion of the lower strut during shimmy, the side dampers will observe a stroke and generate a resistive torque. However, the UTL beam is only free to rotate a few degrees before it engages with the UTL base, which is needed to transfer the torque during a steering operation. Key advantages of STLD are being retrofittable to existing NLGs without the need for a dedicated mount, as well as distributing the mass and loads symmetrically, both of which are shortcomings of existing shimmy dampers [25].

Here, the focus is on development of the UTL structure using topology optimization and conventional design iterations, as elaborated in more detail in a previous publication [12]. Topology optimization based on the SIMP method is employed which is implemented in the Altair OptiStruct™ 2018 numerical solver [15] with the Altair HyperWorks™ 2018 tool being used for creating the FE model. The optimization performed provides a layout of the structure, which is subsequently refined as interpreted by the designer using a CAD package. Since the load applied is based on a shimmy scenario which is dynamic in nature,

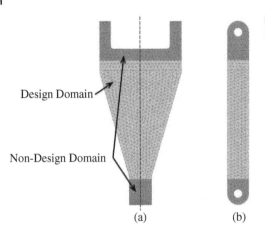

Figure 5.5 Top view (a) and side view (b) of UTL design and non-design domains.

Design Domain

Non-Design Domain

(a)   (b)

equivalent static loads required in the topology optimization are calculated based on the ESL approach implemented in the OptiStruct code. The load profile assumed is [11, 12]:

$$\theta_z(t) = Ae^{-\zeta(2\pi f)t}\sin(2\pi ft) \tag{5.4}$$

where t represents time, $A = 1.1°$ is the shimmy amplitude, $\zeta = 0.01$ is the damping ratio, and $f = 30$ Hz is the shimmy frequency. The parameter values are selected based on Wong et al. [11] while adjusted considering the NLG type assumed here. A worst-case scenario is assumed in selection of the shimmy frequency, since 30 Hz is reported as the higher limit for observed shimmy on landing gears. A transient MBD analysis is incorporated for obtaining the responses of the system, which has the torque links as flexible bodies while the struts and connecting pins are assumed to be rigid. The material properties attributed to the torque links and struts are 7075-T73 Aluminum and 15-5PH Stainless Steel, respectively. The dynamic load is applied to the bottom of the lower strut as a prescribed displacement in the MBD analysis, simulating the shimmy movement of the wheel.

The UTL design domain shown in Figure 5.5 is created based on the available space in the torque link area, while the shape of the mounting lug is preserved as non-design domains according to the baseline system. This is the case since STLD replaces the existing torque links system, adding the functionality of vibration damping to the original purpose of torque transfer. The design space is symmetric since the proposed STLD concept is inherently symmetric to ensure mass, load, and flow balance in and around the NLG. The lower torque link (LTL) is not part of the design domain in this optimization, although it has weight saving potential.

Several topology optimization runs with various parameters are performed. Here three of them are reported which differ in the minimum member size parameter prescribed while the effect of target volume fraction has been investigated previously [12]. The present cases are selected since the minimum member size is observed to have a key influence in the suggested material layouts. The details of the runs are summarized in Table 5.1 and optimization outputs are shown in Figure 5.6. A density threshold of 0.15 is selected to show the topology outcomes, indicating that elements with virtual density values of $\rho_i \geq 0.15$ are interpreted as solid material. For all three cases reported the constraints applied include a target volume fraction of 0.3, global stress limit of 447 MPa, and a one-plane symmetry about

Table 5.1 Topology optimization parameters and run histories.

Run ID	Min. member size (mm)	Total iteration count	Total ESL outer loops
Run 1	None	494	12
Run 2	10	366	11
Run 3	20	283	7

the torque link system's plane of symmetry. As observed, all three runs suggest a V-shaped UTL structure with differences in the wall thickness and internal braces. Without a constraint on the minimum member size, a network of slender members appears inside the design space as seen in the Run 1 layout. These slender members are susceptible to buckling while also increasing manufacturing effort required. The layouts of Run 2 and Run 3 differ only in terms of the thickness of side beams and connection area to the base. Here Run 3 is selected as the design candidate and is reconstructed in CAD for performance analysis as seen in Figure 5.7a. The performance of this design candidate and further iterations on that follow this paragraph. It is also noted that a higher minimum member size has reduced the number of iterations and ESL loops, which translates to lower computational cost here.

The UTL designed based on topology optimization is analyzed and its performance is summarized in Figures 5.7a representing the stress distribution and in Figure 5.8a showing the estimated stroke for side dampers. The stroke is estimated by measuring the relative displacement between the apex joint and the side of the UTL base. This measure overestimates the stroke as less space will be available for the side dampers since mounting lugs need to be added to the UTL base and beam. However, it serves the purpose of comparing design iterations here. One can observe that iteration (a) shows high stress values beyond the endurance limit of the material which is 150 MPa. Hence, it will face fatigue damage. Furthermore, the stroke predicted is less than 0.1 mm which will lead to negligible energy dissipation in the damping elements.

To address these shortcomings, several design iterations are performed on the obtained UTL topology. Here only three of the cases are presented as an illustration of the procedure followed to increase the stroke values. As shown in Figures 5.7 and 5.8, from iteration (b) to (d) the cantilever part of the beam is elongated, which has resulted in increasing the stroke by more than 100%. However, the peak stress remains above the endurance limit in all these cases and new high stress areas emerge on the side beams as a result of this change.

Upon performing additional iterations with the aim of reducing the peak stress, it was observed that they will lead to reduction of the stroke level. This suggests that the two objectives of low compliance (i.e. high stroke) and high strength are conflicting, and a single-piece metallic design may not be possible. Hence, the STLD concept was updated to a two-piece mechanism which comprises a UTL beam and UTL base, connected through a revolute joint as seen in Figure 5.9. Through including a small rotary gap of α which is set to 3° in the presented design, the UTL beam is able to oscillate rotationally about the revolute joint without inducing high stresses in the UTL base. This oscillation which occurs during shimmy will be attenuated by the side dampers mounted symmetrically on the UTL beam and base through revolute joints. As seen in Figure 5.10, the predicted stresses for

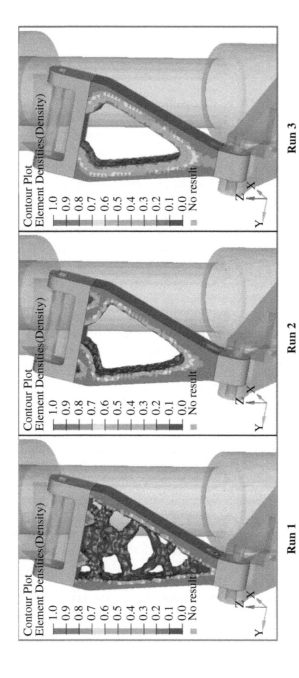

Figure 5.6 Topology optimization results for the UTL using different minimum member size constraints.

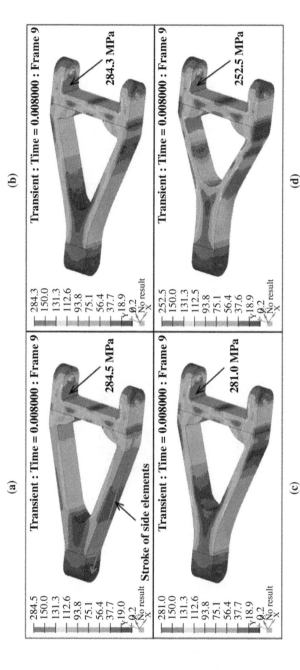

Figure 5.7 von Mises stress contours for the V-shaped UTL design iterations (a)–(d).

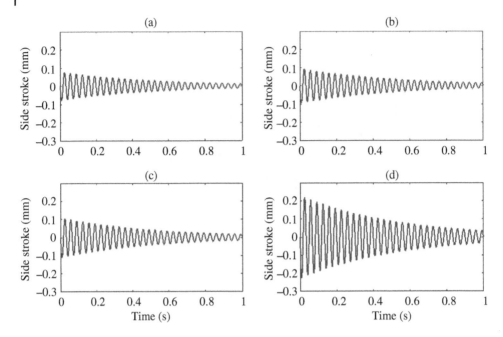

Figure 5.8 Predicted stroke of side dampers for the V-shaped UTL design iterations (a)–(d).

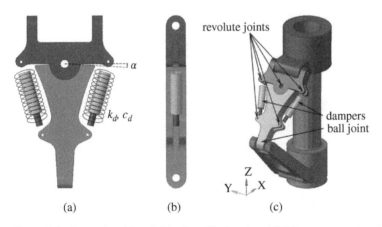

Figure 5.9 Front view (a) and side view (b) of updated STLD concept, and system as mounted on the NLG struts (c).

the updated STLD are all below the endurance strength of 150 MPa. The current design allows for close to 1 mm of stroke for the side dampers as depicted in Figure 5.11, which is roughly 5 times higher than the stroke achievable by the initial STLD concept. Here, the STLD damping coefficient and spring stiffness of $c_d = 15000$ Ns/m and $k_d = 50000$ N/m are incorporated based on parametric studies performed earlier [25, 27]. Although not included here, a frequency response analysis performed on the final design indicated that the resonance frequencies of the system are above 350 Hz, which is well separated from typical shimmy range [12].

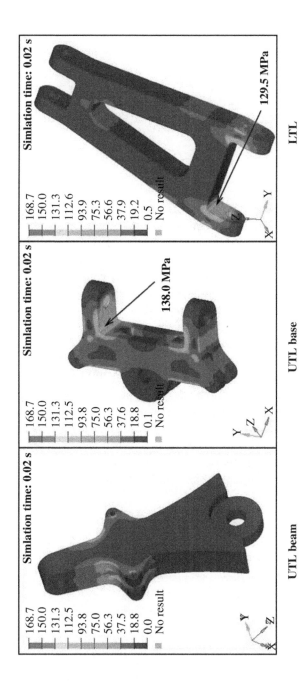

Figure 5.10 von Mises stress contours for the updated STLD concept.

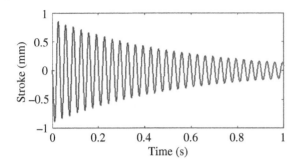

Figure 5.11 Predicted stroke of side dampers for the updated STLD concept.

5.5 Conclusions

An overview of multifunctional aerospace structures is presented in this chapter with emphasis on the computational design and optimization aspect of their development. The case of a novel shimmy damper mechanism for NLGs is elaborated on as an application of the design optimization methods increasingly used in aerospace structures. This novel shimmy damper designated as STLD is proposed to combat the undesired shimmy oscillations observed on some landing gears. This is done through dissipating kinetic energy by the viscous dampers installed symmetrically on the sides of the torque links. The core advantage of STLD is being retrofittable to existing NLGs through integration of the damping mechanism into the structural members of the torque link, hence embodying a multifunctional approach to the design.

Using topology optimization, a V-shape layout is found for the UTL which is subsequently analyzed through a representative MBD model. Upon pursing design iterations to improve the performance, it was concluded that the concept needs to be updated to a two-piece UTL to accommodate the higher stroke needed for the side dampers while ensuring structural integrity under cyclic shimmy loads. The case presented demonstrates the vital role of multifunctional structures in making aerospace systems more compact and easier to maintain, while the application of topology optimization utilized in the process is also revealed.

References

1 Bajpai, R.P., Chandrasekhar, U., and Arankalle, A.R. (2017). *Innovative Design and Development Practices in Aerospace and Automotive Engineering*. Springer. https://doi .org/10.1007/978-981-10-1771-1.

2 Chopra, I. and Sirohi, J. (2013). *Smart Structures Theory*. Cambridge University Press. https://doi.org/10.1017/CBO9781139025164.

3 Sairajan, K.K., Aglietti, G.S., and Mani, K.M. (2016). A review of multifunctional structure technology for aerospace applications. *Acta Astronaut.* 120: 30–42.

4 Seong Jang, T., Soo Oh, D., Kyu Kim, J. et al. (2011). Development of multi-functional composite structures with embedded electronics for space application. *Acta Astronaut.* 68: 240–252.

5 Pereira, T., Guo, Z., Nieh, S. et al. (2008). Embedding thin-film lithium energy cells in structural composites. *Compos. Sci. Technol.* 68: 1935–1941.

6 Adam, T., Liao, G., Petersen, J. et al. (2018). Multifunctional composites for future energy storage in aerospace structures. *Energies* 11: 335.

7 Bendsøe, M.P. and Kikuchi, N. (1988). Generating optimal topologies in structural design using a homogenization method. *Comput. Meth. Appl. Mech. Eng.* 71: 197–224.

8 Munk, D.J., Auld, D.J., Steven, G.P., and Vio, G.A. (2019). On the benefits of applying topology optimization to structural design of aircraft components. *Struct. Multidisc. Optim.* 60: 1245–1266.

9 Bendsøe, M.P. and Sigmund, O. (2004). *Topology Optimization*. Springer. https://doi.org/10.1007/978-3-662-05086-6.

10 Bendsøe, M.P. and Sigmund, O. (1999). Material interpolation schemes in topology optimization. *Arch. Appl. Mech. (Ingenieur Arch.)* 69: 635–654.

11 Wong, J., Ryan, L., and Kim, I.Y. (2017). Design optimization of aircraft landing gear assembly under dynamic loading. *Struct. Multidisc. Optim.* 57: 1357–1375.

12 Rahmani, M. and Behdinan, K. (2020). Structural design and optimization of a novel shimmy damper for nose landing gears. *Struct. Multidisc. Optim.*: 1–36.

13 Ahmad, Z., Sultan, T., Zoppi, M. et al. (2017). Nonlinear response topology optimization using equivalent static loads-case studies. *Eng. Optim.* 49: 252–268.

14 Park, G.J. (2011). Technical overview of the equivalent static loads method for non-linear static response structural optimization. *Struct. Multidisc. Optim.* 43: 319–337.

15 Altair Engineering Inc. (2018). Altair OptiStruct 2018 Reference Guide.

16 Zhou, M., Shyy, Y.K., and Thomas, H.L. (2001). Checkerboard and minimum member size control in topology optimization. *Struct. Multidisc. Optim.* 21: 152–158.

17 Jeong, S.H., Park, S.H., Choi, D.-H., and Yoon, G.H. (2013). Toward a stress-based topology optimization procedure with indirect calculation of internal finite element information. *Comput. Math. Appl.* 66: 1065–1081.

18 Picelli, R., Townsend, S., Brampton, C. et al. (2018). Stress-based shape and topology optimization with the level set method. *Comput. Meth. Appl. Mech. Eng.* 329: 1–23.

19 Remouchamps, A., Bruyneel, M., Fleury, C., and Grihon, S. (2011). Application of a bi-level scheme including topology optimization to the design of an aircraft pylon. *Struct. Multidisc. Optim.* 44: 739–750.

20 Wang, Y., Gao, J., Luo, Z. et al. (2017). Level-set topology optimization for multimaterial and multifunctional mechanical metamaterials. *Eng. Optim.* 49: 22–42.

21 Panesar, A., Ashcroft, I., Brackett, D. et al. (2017). Design framework for multifunctional additive manufacturing: Coupled optimization strategy for structures with embedded functional systems. *Addit. Manuf.* 16: 98–106.

22 Bielefeldt, B.R., Reich, G.W., Beran, P.S., and Hartl, D.J. (2019). Development and validation of a genetic L-System programming framework for topology optimization of multifunctional structures. *Comput. Struct.* 218: 152–169.

23 Mativo, J. and Hallinan, K. (2019). Development of compliant Thermoelectric Generators (TEGs) in aerospace applications using topology optimization. *Energy Harvest Syst.* 4: 87–105.

24 Seifert, R., Patil, M., Seidel, G., and Reich, G. (2019). Multifunctional topology optimization of strain-sensing nanocomposite beam structures. *Struct. Multidisc. Optim.* 60: 1407–1422.

25 Rahmani, M. and Behdinan, K. (2019). On the effectiveness of shimmy dampers in stabilizing nose landing gears. *Aerosp. Sci. Technol.* 91: 272–286.

26 Rahmani, M. and Behdinan, K. (2019). Performance analysis and parametric studies of nose landing gear shimmy dampers. In: *ASME IMECE 2019*. Salt Lake City, UT, USA (8–14 November 2019), 1–6. ASME.

27 Rahmani, M. and Behdinan, K. (2019). On the analysis of passive vibration mitigation of nose landing gears. International Conference on Structural Engineering, Mechanics and Computation (SEMC 2019), Cape Town.

28 Rahmani, M. and Behdinan, K. (2019). Parametric study of a novel nose landing gear shimmy damper concept. *J. Sound Vib.* 457: 299–313.

29 Rahmani, M. and Behdinan, K. (2020). Interaction of torque link freeplay and Coulomb friction nonlinearities in nose landing gear shimmy scenarios. *Int. J. Non Linear Mech.* 119: 103338.

30 Krüger, W.R. and Morandini, M. (2011). Recent developments at the numerical simulation of landing gear dynamics. *CEAS Aeronaut. J.* 1: 55–68.

31 Krüger, W., Besselink, I.J.M., Cowling, D. et al. (1997). Aircraft landing gear dynamics: simulation and control. *Veh. Syst. Dyn.* 28: 119–158.

32 Eret, P., Kennedy, J., and Bennett, G.J. (2015). Effect of noise reducing components on nose landing gear stability for a mid-size aircraft coupled with vortex shedding and freeplay. *J. Sound Vib.* 354: 91–103.

33 Howcroft, C., Lowenberg, M., Neild, S. et al. (2015). Shimmy of an aircraft main landing gear with geometric coupling and mechanical freeplay. *J. Comput. Nonlinear Dyn.* 10: 051011.

34 Rahmani, M. and Behdinan, K. (2019). Investigation on the effect of coulomb friction on nose landing gear shimmy. *J. Vib. Control* 25: 255–272.

35 Thota, P., Krauskopf, B., Lowenberg, M., and Coetzee, E. (2010). Influence of tire inflation pressure on nose landing gear shimmy. *J. Aircr.* 47: 1697–1706.

36 Tourajizadeh, H. and Zare, S. (2016). Robust and optimal control of shimmy vibration in aircraft nose landing gear. *Aerosp. Sci. Technol.* 50: 1–14.

37 Atabay, E. and Ozkol, I. (2013). Application of a magnetorheological damper modeled using the current-dependent Bouc-Wen model for shimmy suppression in a torsional nose landing gear with and without freeplay. *J. Vib. Control* 20: 1622–1644.

38 Arreaza, C., Behdinan, K., and Zu, J.W. (2016). Linear stability analysis and dynamic response of shimmy dampers for main landing gears. *J. Appl. Mech.* 83: 081002.

6

Dynamic Modeling and Analysis of Nonlinear Flexible Rotors Supported by Viscoelastic Bearings

Mohammed Khair Al-Solihat and Kamran Behdinan

Advanced Research Laboratory for Multifunctional Lightweight Structures (ARL-MLS), Department of Mechanical & Industrial Engineering, University of Toronto, Toronto, Canada

6.1 Introduction

Rotors constitute a primary part of turbomachines such as pumps, compressors, and jet engines [1–3]. To analyze the dynamic behavior of rotor systems, simple rigid body dynamic models of rotors are first developed. Rigid body models of rotors, often referred to as *Jeffcott rotor* models, consider the rotor system as a rigid disk or cylinder rotating around its axis and supported on viscoelastic elements representing the combined shaft and bearing stiffness and damping [4, 5]. However, flexible models of rotors consider the shaft elasticity and corresponding geometric nonlinearities. Thus, these models are more efficient to predict the nonlinear behavior of the system arising from the nonlinear elastic motion of the shaft [3, 6–8].

Primary flexible rotor models are developed for rotating shafts (without disk) to study the free vibration characteristics. Pai et al. [7] constructed the Campbell diagrams of a rotating shaft (modeled as Rayleigh's beam) with different boundary conditions. The effects of gravity and slenderness ratio were also investigated. This model corrected the formulation of these earlier models [9, 10]. Similar models for simply supported extensible and inextensible rotating shafts were also developed by Khadem and his co-workers [11, 12] to study the nonlinear response to the shaft unbalance force. The disk(s) dynamics were then incorporated into the flexible rotating shaft models to analyze the dynamic behavior due to unbalance disk and rotor-stator contact forces. Khanlo et al. [13] analyzed the dynamic behavior of a simply supported shaft–disk rotor experiencing rub-impact loads. Tiaki et al. [14] developed a flexible model of an overhung rotor system and analyzed the nonlinear response to disk unbalance and skew using the method of multiple scales. Following the same approach, Shad et al. [15] determined the nonlinear frequency response of a shaft–disk system to unbalance disk force. Utilizing Timoshenko beam theory, a flexible dynamic model with consideration of the supports stiffness nonlinearities for a shaft–disk rotor supported by viscoelastic supports was developed by Shabaneh and Zu [16, 17]. Finally, Phadatare et al. [3, 18] incorporated the effect of the base motion to analyze the free vibration characteristics and nonlinear frequency response of the rotor system.

Advanced Multifunctional Lightweight Aerostructures: Design, Development, and Implementation,
First Edition. Kamran Behdinan and Rasool Moradi-Dastjerdi.
© 2021 John Wiley & Sons Ltd. This Work is a co-publication between John Wiley & Sons Ltd and ASME Press.

Rotors are supported by a suspension system comprising rigid or hydrodynamic bearings. The stiffness and damping properties of the bearings considerably affect the dynamic behavior of the system and the isolation efficiency of the suspension system. Recently, many works have focused on the dynamic behavior and vibration isolation performance of a flexible rotor mounted on flexible bearings or supports. Ribeiro et al. [19, 20] used the finite element method (FEM) to optimize the vibration isolation performance of a flexible shaft–disk rotor supported on viscoelastic supports. Similarly, Sinou [21] used FEM to analyze the nonlinear dynamics of a flexible rotor mounted on ball bearings with consideration of the nonlinear stiffness of the bearings. Bab et al. [22] investigated the isolation performance of nonlinear energy sinks mounted on disk and the bearing of the rotor. The rotor system was supported on hydrodynamic bearing with nonlinear stiffness while the bearing damping was assumed linear. Most recently, Li et al. [23] investigated the effects of bearing stiffness nonlinearity on the nonlinear response of a rotor-blade system using the method of multiple scales.

Most studies discussing the dynamic behavior and vibration isolation performance of rotors supported on viscoelastic bearings/supports are based on simplified rigid body models while few use the flexible rotor model. In this regard, we will present a detailed formulation to model flexible shaft–disk rotor systems. The dynamic modeling and solution techniques for the equations of motion presented herein can be used by future investigators for further investigations on the system dynamics and vibration isolation performance of the suspension systems of rotors.

6.2 Dynamic Modeling

A schematic diagram of the shaft–disk rotor system to be analyzed is illustrated in Figure 6.1. The system components include a rigid disk and flexible shaft (of length l) supported by flexible bearings. The shaft is considered as an Euler–Bernoulli beam supporting a rigid disk at distance, x_D from the left end of the shaft.

The reference frames, XYZ and xyz are introduced to characterize the system kinematics. The XYZ is the inertial frame where the Z axis is assumed to be in the opposite direction of the acceleration of gravity. The xyz frame is attached to the elastic axis of the shaft, as shown in Figure 6.1. The bending deformations of the shaft's elastic axis are denoted by $v(x, t)$ and $w(x, t)$ along the Y and Z axes, respectively, and $u(x, t)$ represents the longitudinal displacement along the X axis.

The restoring force of the bearing due to deflections of the shaft ends is [24]

$$F_{s,ij} = k_{ij}\delta_{ij} + \overline{k}_{ij}\delta_{ij}^3 \tag{6.1}$$

Similarly, the bearing damping force is [24]

$$F_{d,ij} = c_{ij}\dot{\delta}_{ij} + \overline{c}_{ij}\dot{\delta}_{ij}^3 \tag{6.2}$$

where the linear and cubic stiffness coefficients of the bearings are denoted by k_{ij} and $\overline{k}_{ij}$, respectively. Also, c_{ij} and $\overline{c}_{ij}$ represent the linear and cubic damping coefficients, respectively. The subscript $i = Y, Z$ represents directions along Y or Z and $j = L, R$ denotes the left and right sides, respectively. The spinning speed of the shaft (Ω) is assumed constant

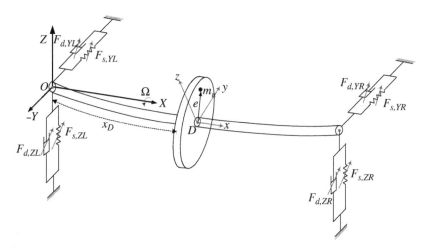

Figure 6.1 Flexible shaft–disk rotor system supported on viscoelastic bearings.

around the X axis and the spinning angle can be expressed as $\beta = \Omega t$ where t is the time in seconds. The system is excited by harmonic force caused by the disk unbalance mass (m_u) at radius e from the disk center, as shown in Figure 6.1. The displacement and velocity of the shaft end are denoted by δ_{ij} and $\dot{\delta}_{ij}$, respectively.

The equations of motion for the system are derived using Lagrange's equation such that the displacement and velocity of shaft elastic axis, and rotation of the shaft cross section should be formulated.

The angular velocity of the shaft cross section, ω_p expressed in the body frame (xyz) can be expressed as [24]

$$^{xyz}\omega_p = \begin{bmatrix} \omega_x \\ \omega_y \\ \omega_z \end{bmatrix} \approx \begin{bmatrix} \Omega + \dot{v}'w' \\ \dot{v}' \sin(\beta) - \dot{w}' \cos(\beta) \\ \dot{v}' \cos(\beta) + \dot{w}' \sin(\beta) \end{bmatrix} \tag{6.3}$$

and in the inertial frame (XYZ) as

$$^{XYZ}\omega_p \approx \begin{bmatrix} \Omega + \dot{w}'v' \\ -\dot{w}' + \Omega v' \\ \dot{v}' + \Omega w' \end{bmatrix} \tag{6.4}$$

The velocity of the unbalance mass can be obtained in the inertial frame as

$$\mathbf{v}_{un} = \mathbf{v}_D + {}^{XYZ}\omega_D \times \mathbf{r}_{un} \tag{6.5}$$

where $\mathbf{v}_D = \dot{\mathbf{u}}|_{x=x_D}$ denotes the velocity of the disk center and $\dot{\mathbf{u}} = [\dot{u}\ \dot{v}\ \dot{w}]^T$ is the velocity of an arbitrary point on the elastic axis of the shaft. The angular velocity of the disk is $\omega_D = \omega_p|_{x=x_D}$. The displacement of the unbalance mass relative to the disk center ($\mathbf{r}_{un}$) can be written in the XYZ frame as

$$\mathbf{r}_{un} \simeq e\left[(v' \sin \beta - w' \cos \beta) \quad -\sin \beta \quad \cos \beta\right]^T \tag{6.6}$$

6.2.1 Equations of Motion and Method of Solution

The kinetic energy of the shaft can be expressed as [4]

$$T_s = \frac{1}{2} \int_0^l \left[m_s \left(\dot{u}^2 + \dot{v}^2 + \dot{w}^2 \right) + \rho I_d \left(\omega_y^2 + \omega_z^2 \right) + \rho I_p \omega_x^2 \right] dx \tag{6.7}$$

where ρ is the shaft density, m_s is the shaft mass per length, and I_p and I_d are the polar and diametral second moment of inertia of the shaft cross section, respectively. The kinetic energy of the disk can be formulated as

$$T_D = \frac{1}{2} \left[M_D \left(\dot{u}^2 + \dot{v}^2 + \dot{w}^2 \right) + I_{Dd} \left(\omega_y^2 + \omega_z^2 \right) + I_{Dp} \omega_x^2 \right]_{x=a} \tag{6.8}$$

where the disk mass is m_D, and I_{Dp} and I_{Dd} denote the polar and diametral mass moment of inertia of the disk around its principal axes, respectively. The kinetic energy of the unbalance mass can be written as

$$T_{un} = \frac{1}{2} m_u \mathbf{v}_{un}^T \mathbf{v}_{un} \tag{6.9}$$

The shaft strain energy can be written as

$$V_s = \frac{1}{2} \int_0^l EI_d \left[v''^2 + w''^2 \right] dx + \frac{1}{2} \int_0^l EA \left[u' + \frac{1}{2} \left(v'^2 + w'^2 \right) \right]^2 dx$$

$$+ \int_0^l \left[\frac{1}{2} \left(k_{YL} v^2 + k_{ZL} w^2 \right) + \frac{1}{4} \left(\bar{k}_{YL} v^4 + \bar{k}_{ZL} w^4 \right) \right] \delta(x) \, dx$$

$$+ \int_0^l \left[\frac{1}{2} \left(k_{YR} v^2 + k_{ZR} w^2 \right) + \frac{1}{4} \left(\bar{k}_{YR} v^4 + \bar{k}_{ZR} w^4 \right) \right] \delta(x - l) \, dx \tag{6.10}$$

where E is Young's modulus, A is the shaft cross-sectional area, and $\delta(.)$ is the Dirac delta function of $(.)$. Rayleigh's dissipation function characterizing the energy dissipated by the bearing damping, can be expressed as [25, 26]

$$\Gamma = \int_0^l \left[\frac{1}{2} \left(c_{YL} \dot{v}^2 + c_{ZL} \dot{w}^2 \right) + \frac{1}{4} \left(\bar{c}_{YL} \dot{v}^4 + \bar{c}_{ZL} \dot{w}^4 \right) \right] \delta(x) \, dx$$

$$+ \int_0^l \left[\frac{1}{2} \left(c_{YR} \dot{v}^2 + c_{ZR} \dot{w}^2 \right) + \frac{1}{4} \left(\bar{c}_{YR} \dot{v}^4 + \bar{c}_{ZR} \dot{w}^4 \right) \right] \delta(x - l) \, dx \tag{6.11}$$

Utilizing the assumed mode method, u, v, and w can be approximately expressed as [27]

$$u(x, t) = \sum_{i=1}^{n_u} \Psi_i(x) \, \eta_i(t)$$

$$v(x, t) = \sum_{i=1}^{n_b} \varphi_i(x) \, q_i(t)$$

$$w(x, t) = \sum_{i=1}^{n_b} \varphi_i(x) \, Q_i(t) \tag{6.12}$$

where $\varphi_i(x)$ and $\Psi_i(x)$ are appropriate shape functions for the bending and axial deflections, respectively. The number of mode shapes for bending and axial deflections of the shaft are n_b and n_u, respectively. The time variables, $\eta_i(t)$, $q_i(t)$, and $Q_i(t)$, are generalized coordinates to be determined by solving the equations of motion.

The eigenfunctions of a free-free beam are chosen for φ_i as [28, 29]

$$\varphi_i = \begin{cases} 1 & \text{for } i = 1 \\ 2x - 1 & \text{for } i = 2 \\ \cosh\left(\lambda_i x\right) + \cos\left(\lambda_i x\right) - \gamma_i \left[\sinh\left(\lambda_i x\right) + \sin\left(\lambda_i x\right)\right] & \text{for } i \geq 3 \end{cases} \tag{6.13}$$

where

$$\gamma_i = \frac{\cos \lambda_i - \cosh \lambda_i}{\sin \lambda_i - \sinh \lambda_i} \tag{6.14}$$

The first two modes (for $i=1$ and 2) represent the rigid body motion in translation and rotation ($\lambda_1 = \lambda_2 = 0$). The remaining modes correspond to the elastic motion of the shaft. The eigenvalue (λ_i when $i \geq 3$) can be determined by solving the transcendental relation, $\cos\lambda_i \cosh \lambda_i = 1$ [27]. In addition, the eigenfunctions of a clamped-free longitudinal vibration of a rod [27] can be used for Ψ_i.

As demonstrated earlier, the equations of motion are obtained using Lagrange's equation as [26]

$$\frac{d}{dt}\left(\frac{\partial \mathcal{L}}{\partial \dot{\chi}}\right) - \frac{\partial \mathcal{L}}{\partial \chi} + \frac{\partial \Gamma}{\partial \dot{\chi}} = 0 \tag{6.15}$$

where $\mathcal{L} = \mathcal{T} - \mathcal{V}$ represents the Lagrangian, $\mathcal{T} = \mathcal{T}_s + \mathcal{T}_D + \mathcal{T}_{un}$, and $\mathcal{V} = \mathcal{V}s + \mathcal{V}_g$, respectively, and $\chi = [\mathbf{q} \quad \mathbf{Q} \quad \eta]^T$ denotes the generalized coordinates vector. The equations of motion can expressed in the following form

$$\mathbf{M}\ddot{\chi} = \mathbf{F} \tag{6.16}$$

where $\mathbf{M}$ is the mass matrix and $\mathbf{F}$ is the right-hand side of the equations of motion which is nonlinear in $\chi, \dot{\chi}, \Omega$ and t. The following dimensionless parameters are introduced to avoid numerical instability

$$x^* = \frac{x}{l} \quad , \quad v^* = \frac{v}{l} \quad , \quad w^* = \frac{w}{l} \quad , \quad u^* = \frac{u}{l} \quad , \quad x_D^* = \frac{x_D}{l} \quad ,$$

$$\omega_0 = \left(\frac{\pi}{l}\right)\sqrt{\frac{E}{\rho}} \quad , \quad \tau = \omega_0 t \quad , \quad \Omega^* = \frac{\Omega}{\omega_0}, \quad k_{(ij)}^* = \frac{k_{(ij)} l^3}{EI_d} \quad , \quad c_{(ij)}^* = \frac{c_{(ij)}}{m_s l \omega_0}$$

$$\bar{k}_{(ij)}^* = \frac{\bar{k}_{(ij)} l^5}{EI_d} \quad , \quad \bar{c}_{(ij)}^* = \frac{\bar{c}_{(ij)} \omega_0 l}{m_s} \tag{6.17}$$

substituting Eq. (6.17) into Eq. (6.16) will render it in a dimensionless form.

6.2.2 Force Transmissibility

The total force transmitted to the bearings due to restoring and damping forces of the bearings can be formulated as

$$\mathbf{F}_T = \mathbf{F}_{s,YR} + \mathbf{F}_{s,YL} + \mathbf{F}_{s,ZR} + \mathbf{F}_{s,ZL} + \mathbf{F}_{d,YR} + \mathbf{F}_{d,YL} + \mathbf{F}_{d,ZR} + \mathbf{F}_{d,ZL} \tag{6.18}$$

The force transmissibility can be defined as the ratio of the total transmitted force to the bearing (or foundation) to the unbalance force, formulated as

$$T_F = \max\left(\frac{\|\mathbf{F}_T\|}{m_u e \Omega^2}\right) \tag{6.19}$$

where $\|\mathbf{F}_T\|$ is the magnitude of $\mathbf{F}_T$.

6.2.3 Method of Solution

Linear equations of motion have unique solutions which can be solved analytically, often numerically. Explicit numerical integration techniques such as Newton's, finite difference and Runge–Kutta methods can be used to integrate the differential equations of motion [30]. Also, implicit numerical integration techniques such as Newmark-β and Generalized-α are used for the same purpose and they are widely used, particularly in structural dynamics applications [31]. However, when the equations of motion are nonlinear, as the rotor system under consideration, the equations of motion have multiple solutions such that different sets of initial conditions could yield different solutions. Therefore, perturbation methods such as the methods of multiple scales, averaging and harmonic balance (HB), are widely used to solve the nonlinear equations of motion [32]. The latter approach will be adopted in this work because it is simpler and more efficient to obtain the frequency response particularly for multiple resonance frequencies. This approach (HB) assumes the solution as the sum of many harmonic terms (sines and cosines with different coefficients), as [32]

$$\chi = \mathbf{a}_0 + \sum_{i=1}^{n_H} \left[\mathbf{A}_i \sin\left(i\Omega\tau\right) + \mathbf{B}_i \cos\left(i\Omega\tau\right) \right] \tag{6.20}$$

where n_H denotes the number of harmonic terms and $\mathbf{a}_0$ is a constant vector to account for the mean response. The coefficients of the sine and cosine terms are $\mathbf{A}_i$ and $\mathbf{B}_i$, respectively.

Equation (6.20) can be then substituted into the equations of motion (Eq. (6.16)) to obtain a system of $n \times (2n_H + 1)$ nonlinear algebraic equations, formulated as

$$\mathbf{f}(\mathbf{X}, \Omega) = 0 \quad , \quad \mathbf{X} = \left[\mathbf{a}_0, \mathbf{A}_k, \mathbf{B}_k \right] \tag{6.21}$$

where $n = 2n_b + n_u$. A continuation scheme can be subsequently used to construct the frequency response curves, i.e. $\Omega - \mathbf{X}$ relationship.

6.3 Free Vibration Characteristics

The free vibration analysis can be performed by linearizing the equations of motion and neglecting the external force terms. The eigen value problem can be then solved to calculate the natural frequencies (ω_n) of the system and corresponding whirling modes. Thus, so-called Campbell diagrams that trace the $\Omega - \omega_n$ relationship for different mode shapes (whirling modes) can be constructed. It is important to know that when the whirling motion of the shaft follows the same direction of the shaft spin speed (Ω) it is defined as a *forward whirl* (FW) while a *backward whirl* (BW) represents a whirling motion in the opposite direction of Ω.

Figure 6.2 shows the Campbell diagram of the shaft–disk rotor system with identical stiffness at both the shaft ends ($k = 200$) and with disk location $x_D = \frac{1}{6}$. The points of intersection of the synchronous whirl line ($\Omega = \omega_n$) with the whirling modes represent the critical speeds. It is clear that the forward and backward whirling modes of the first and second modes are almost equal and do not vary with the spinning speed increment up to certain speed and then split where the forward (FW) modes keep increasing with Ω increase while the backward (BW) modes decrease as Ω increases. However, the natural frequency of the

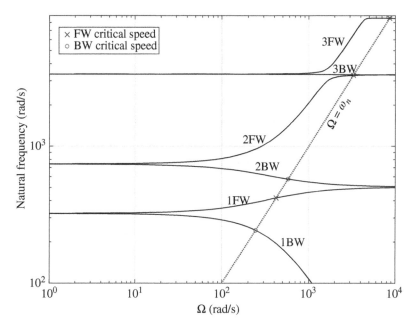

Figure 6.2 Campbell diagram showing the forward and backward whirling modes and their corresponding critical speeds for $x_D = \frac{1}{6}$ and $k = 200$. Source: The shaft–disk properties are based on [4].

third backward mode (3BW) is almost insensitive to Ω variations while the corresponding forward mode (3FW) splits and proceed upwards, i.e. increases as Ω increases. It is important to note that the characteristics of the Campbell diagram vary with bearing stiffness (k) and disk location (x_D).

Figure 6.3 illustrates the effects of the bearing stiffness and disk location on the first and second forward critical speeds. Typically, the critical speeds increase as the bearing stiffness increases. However, the corresponding effect of x_D varies among the whirling modes due to the gyroscopic effect of the disk. In general, the closer disk location (lower x_D) yields higher first critical speed at higher bearing stiffness while this effect is almost negligible for lower k, as shown in Figure 6.3a. In contrast, it is evident that lower x_D yields lower second critical speeds, as illustrated in Figure 6.3b.

6.4 Nonlinear Frequency Response

The nonlinear behavior of the flexible rotor system is triggered by the geometric nonlinearities of the shaft flexibility due to shortening and stretching effects. In addition, the nonlinear stiffness and damping properties of the bearings also enhance the nonlinear behavior considerably. In this section, we will perform several dynamic simulations for the unbalanced rotor to investigate the effects of these nonlinearities on the nonlinear dynamic response and force transmissibility of the system.

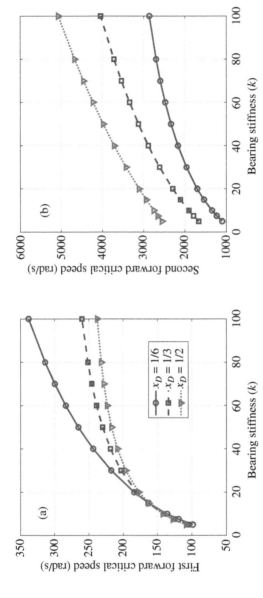

Figure 6.3 The effects of the bearing stiffness and disk location on (a) first forward critical speed and (b) second forward critical speed.

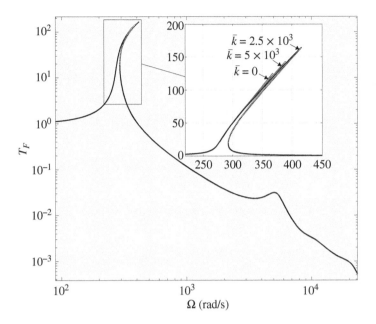

Figure 6.4 Influence of $\bar{k}$ on the force transmissibility at $m_u = 10$ gm, $k = 200$, $c = 0.01$, and $\bar{c} = 0$.

Figures 6.4–6.6 illustrate the effect of the bearing stiffness nonlinearity $(\bar{k})$ on the frequency response and force transmissibility for stiff bearings $(k = 200)$. When the bearing stiffness is linear $(\bar{k} = 0)$, it is clear that the hardening effect and dynamic instability are only visible in the first resonance peak while the remaining peaks are stable. Also, the effect of the bearing nonlinearity is only visible in the vicinity of the first resonance region such that higher $\bar{k}$ yields larger force transmissibility (T_F) and the frequency response of the first and second generalized coordinates of the bending deflections $(q_1$ and $q_2)$. The effect of bearing nonlinearity on the higher order resonance regions is small because the system response (bending deflections) are very small and not sufficient to trigger the nonlinear effect. It is worth noting that the nonlinear behavior of the system when supported by relatively soft bearing is quite different [24].

Figures 6.7–6.9 show the influence of cubic nonlinear damping on the nonlinear frequency response and force transmissibility of the unbalanced shaft–disk rotor system. The results show that the influence of nonlinear damping is more considerable on the higher resonance region where large reductions in the system response and force transmissibility are more evident in the third peak. However, the first resonance peak seems insensitive to $\bar{c}$ while a moderate reduction is visible in the second resonance region. In fact, the nonlinear damping force is proportional to the cube of the spinning speed of the shaft which explains the considerable effect of the nonlinear damping at higher Ω.

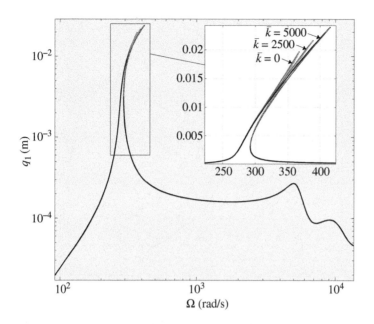

Figure 6.5 Influence of $\bar{k}$ on the frequency response of q_1 at $m_u = 10$ gm, $k = 200$, $c = 0.01$, and $\bar{c} = 0$.

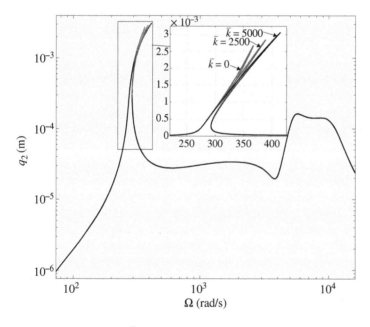

Figure 6.6 Influence of $\bar{k}$ on the frequency response of q_2 at $m_u = 10$ gm, $k = 200$, $c = 0.01$, and $\bar{c} = 0$.

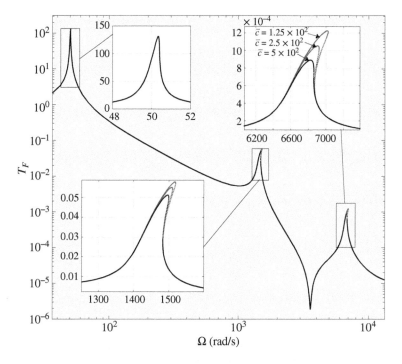

Figure 6.7 Influence of $\bar{c}$ on the force transmissibility at $m_u = 4$ gm, $k = 1$, $\bar{k} = 0$, and $c = 1 \times 10^{-4}$.

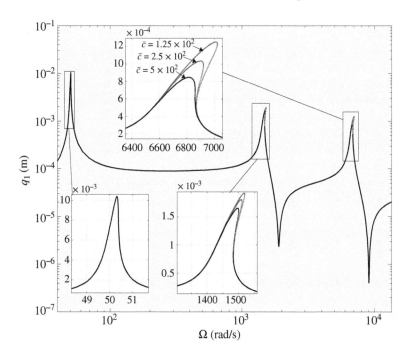

Figure 6.8 Influence of $\bar{c}$ on the frequency response of q_1 at $m_u = 4$ gm, $k = 1$, $\bar{k} = 0$, and $c = 1 \times 10^{-4}$.

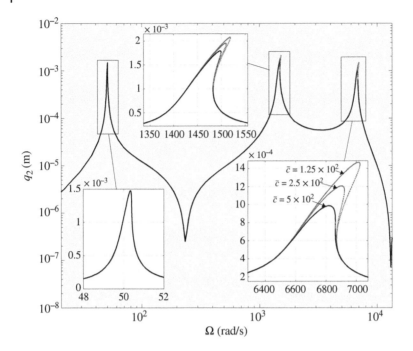

Figure 6.9 Influence of $\bar{c}$ on the frequency response of q_2 at $m_u = 1$ gm, $k = 1$, $\bar{k} = 0$, and $c = 1 \times 10^{-4}$.

6.5 Conclusions

The nonlinear dynamic model developed for a flexible shaft–disk rotor system, supported by a nonlinear suspension system with cubic nonlinearities in stiffness and damping, has been analyzed using the HB method combined with a continuation scheme. The Campbell diagrams have been constructed to determine the effects of the bearing linear stiffness and shaft–disk location on the magnitudes of the critical speeds. It was found that the closer the disk location to the shaft end the higher the first critical speed, while the second critical speed is affected in an adverse manner.

Many numerical simulations were conducted to examine the effects of the cubic bearing stiffness and damping on the nonlinear frequency response and force transmissibility characteristics. For stiff bearings, the simulation results showed that the bearing stiffness nonlinearity leads to higher response and force transmissibility magnitudes at the first resonance region while the remaining regions are not affected. The effect of nonlinear damping was investigated for a soft suspension system ($k = 1$). The results showed that the effect of the nonlinear damping is proportional to the cube of the spinning speed such that the high order resonance regions exhibited higher reduction (as $\bar{c}$ increases) in the magnitudes of the system response and force transmissibility.

References

1 Sun, C., Chen, Y., and Hou, L. (2016). Steady-state response characteristics of a dual-rotor system induced by rub-impact. *Nonlinear Dyn.* 86 (1): 91–105.

2 Chun, S.-B. and Lee, C.-W. (1996). Vibration analysis of shaft-bladed disk system by using substructure synthesis and assumed modes method. *J. Sound Vib.* 189 (5): 587–608.

3 Phadatare, H., Choudhary, B., and Pratiher, B. (2017). Evaluation of nonlinear responses and bifurcation of a rotor-bearing system mounted on moving platform. *Nonlinear Dyn.* 90 (1): 493–511.

4 Al-Solihat, M.K. and Behdinan, K. (2019). Nonlinear dynamic response and transmissibility of a flexible rotor system mounted on viscoelastic elements. *Nonlinear Dyn.* 97 (2): 1581–1600.

5 Zou, D., Rao, Z., and Ta, N. (2015). Coupled longitudinal-transverse dynamics of a marine propulsion shafting under superharmonic resonances. *J. Sound Vib.* 346: 248–264.

6 Yu, T.-J., Zhou, S., Yang, X.-D., and Zhang, W. (2018). Global dynamics of a flexible asymmetrical rotor. *Nonlinear Dyn.* 91 (2): 1041–1060.

7 Pai, P.F., Qian, X., and Du, X. (2013). Modeling and dynamic characteristics of spinning rayleigh beams. *Int. J. Mech. Sci.* 68: 291–303.

8 Roques, S., Legrand, M., Cartraud, P. et al. (2010). Modeling of a rotor speed transient response with radial rubbing. *J. Sound Vib.* 329 (5): 527–546.

9 Sheu, G. and Yang, S.-M. (2005). Dynamic analysis of a spinning Rayleigh beam. *Int. J. Mech. Sci.* 47 (2): 157–169.

10 Zu, J.W.-Z. and Han, R.P. (1992). Natural frequencies and normal modes of a spinning Timoshenko beam with general boundary conditions. *J. Appl. Mech.* 59 (2S): S197–S204.

11 Khadem, S., Shahgholi, M., and Hosseini, S. (2010). Primary resonances of a nonlinear in-extensional rotating shaft. *Mech. Mach. Theory* 45 (8): 1067–1081.

12 Shahgholi, M. and Khadem, S. (2012). Primary and parametric resonances of asymmetrical rotating shafts with stretching nonlinearity. *Mech. Mach. Theory* 51: 131–144.

13 Khanlo, H., Ghayour, M., and Ziaei-Rad, S. (2011). Chaotic vibration analysis of rotating, flexible, continuous shaft-disk system with a rub-impact between the disk and the stator. *Commun. Nonlinear Sci. Numer. Simul.* 16 (1): 566–582.

14 Tiaki, M.M., Hosseini, S., and Zamanian, M. (2016). Nonlinear forced vibrations analysis of overhung rotors with unbalanced disk. *Arch. Appl. Mech.* 86 (5): 797–817.

15 Shad, M.R., Michon, G., and Berlioz, A. (2011). Modeling and analysis of nonlinear rotordynamics due to higher order deformations in bending. *Appl. Math. Modell.* 35 (5): 2145–2159.

16 Shabaneh, N. and Zu, J. (2003). Nonlinear dynamic analysis of a rotor shaft system with viscoelastically supported bearings. *J. Vib. Acous. Trans. ASME* 125 (3): 290–298.

17 Ji, Z. and Zu, J. (1998). Method of multiple scales for vibration analysis of rotor-shaft systems with non-linear bearing pedestal model. *J. Sound Vib.* 218 (2): 293–305.

18 Phadatare, H., Maheshwari, V., Vaidya, K., and Pratiher, B. (2017). Large deflection model for nonlinear flexural vibration analysis of a highly flexible rotor-bearing system. *Int. J. Mech. Sci.* 134: 532–544.

19 Ribeiro, E.A., Pereira, J.T., and Bavastri, C.A. (2015). Passive vibration control in rotor dynamics: optimization of composed support using viscoelastic materials. *J. Sound Vib.* 351: 43–56.

20 Ribeiro, E.A., de Oliveira Lopes, E.M., and Bavastri, C.A. (2017). A numerical and experimental study on optimal design of multi-DOF viscoelastic supports for passive vibration control in rotating machinery. *J. Sound Vib.* 411: 346–361.

21 Sinou, J.-J. (2009). Non-linear dynamics and contacts of an unbalanced flexible rotor supported on ball bearings. *Mech. Mach. Theory* 44 (9): 1713–1732.

22 Bab, S., Khadem, S., Shahgholi, M., and Abbasi, A. (2017). Vibration attenuation of a continuous rotor-blisk-journal bearing system employing smooth nonlinear energy sinks. *Mech. Syst. Sig. Process.* 84: 128–157.

23 Li, B., Ma, H., Yu, X. et al. (2019). Nonlinear vibration and dynamic stability analysis of rotor-blade system with nonlinear supports. *Arch. Appl. Mech.*: 1–28.

24 Al-Solihat, M.K. and Behdinan, K. (2020). Force transmissibility and frequency response of a flexible shaft-disk rotor supported by a nonlinear suspension system. *Int. J. Non Linear Mech.* 124: 103501.

25 Meirovitch, L. and Tuzcu, I. (2003). Integrated approach to the dynamics and control of maneuvering flexible aircraft. NASA Langley Research Center, Hampton, VA. *Tech. Rep. NASA/CR-2003-211748.*

26 Meirovitch, L. and Stemple, T. (1995). Hybrid equations of motion for flexible multibody systems using quasicoordinates. *J. Guid. Control Dyn.* 18 (4): 678–688.

27 Rao, S.S. (2007). *Vibration of Continuous Systems.* New York: Wiley.

28 Karnovsky, I.A., Lebed, O.I., and Lebed, O. (2000). *Formulas for Structural Dynamics: Tables, Graphs and Solutions.* New York: McGraw-Hill.

29 Kheiri, M., Païdoussis, M., and Amabili, M. (2013). A nonlinear model for a towed flexible cylinder. *J. Sound Vib.* 332 (7): 1789–1806.

30 Gautschi, W. (1997). *Numerical Analysis.* Springer Science & Business Media.

31 Leondes, C.T. (1999). *Structural Dynamic Systems Computational Techniques and Optimization: Computational Techniques*, vol. 13. CRC Press.

32 Thomsen, J.J. (2013). *Vibrations and Stability: Advanced Theory, Analysis, and Tools.* Heidelberg: Springer Science & Business Media.

7

Modeling and Experimentation of Temperature Calculations for Belt Drive Transmission Systems in the Aviation Industry

Xingchen Liu and Kamran Behdinan

Advanced Research Laboratory for Multifunctional Lightweight Structures (ARL-MLS), Department of Mechanical & Industrial Engineering, University of Toronto, Toronto, Canada

7.1 Introduction

The belt drive system is a common but critical transmission component in helicopter transmission systems [1]. This system depicted in Figure 7.1 consists of driver (DR) and driven (DN) pulleys. It transmits power from the engine to the gearbox of the propeller. Traditionally, pulleys in this type of system are made of steel for high-torque transmissions. However, a novel type of fiber-reinforced polymer (FRP) could potentially be used for pulley fabrication, due to its favorable strength-to-weight ratio. However, its application in the aviation industry is not straightforward. The low thermal conductivity of FRP causes surface deterioration in the pulleys and belt, reducing the life of the system. To improve reliability, surface temperature profiles must be acquired in advance, especially during the design stage. Therefore, in order to avoid this type of failure, this study proposes a thermal model that provides immediate temperature distributions for belt drive system design.

The thermal model presented here is divided into two subanalyses: calculation of heat generation and location in the system, and simulation of thermal flow and dissipation inside the components within the system. Heat is originally generated from power loss inside the system because all lost power is transformed into heat. It then spreads inside the component, and finally dissipates into the surrounding air through the exposed surfaces of each part. Currently, there are several well-established theories related to power loss calculation. Although the original purpose of this research was to enhance belt drive efficiency, this work can also be used to calculate heat generation in belt drive systems. One sophisticated theory identified five different causes for power loss in belt drives: belt bending, shear, radial compression, tension, and slippage against the pulley [2]. Due to the belt's high rotation speed, the first four losses can be treated as belt-inherent power losses, while the last can be considered frictional heat generated at surfaces engaged with both the belt and pulley. This theory was later expanded to incorporate the heavy-duty poly-V belt drive system [3]. Other research had considered slip power loss to be the main contributor, neglecting other types [4]. Recent research modified this theory by assuming that power loss results from bearing frictional loss [5], as well as the hysteresis of rubber deformation attributable to the belt's

Advanced Multifunctional Lightweight Aerostructures: Design, Development, and Implementation,
First Edition. Kamran Behdinan and Rasool Moradi-Dastjerdi.

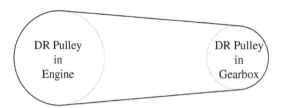

Figure 7.1 Schematic of a belt drive system in a typical helicopter.

dynamic bend, stretch, shear, flank, and radial compression [6, 7]. Bertini et al. proposed that power loss be classified into three types: lateral and longitudinal hysteresis of the belt rubber, frictional sliding at surfaces in contact with the belt and pulley, and frictional losses when the pulley engages/disengages with the belt [8]. Other research indicated that several parameters impact power loss. The first was belt clamp pressure on the outer surface of the pulley. Qin et al. investigated the effects of various frictional clamp forces on relative axial displacement inside a clamp band joint during simulations of hysteresis behavior [9]. Bent showed experimentally that belt pressure influences the axial force of the pulley, along with the overall efficiency of the transmission system [10]. A dynamic model was established by Julio and Plante to calculate the transmission ratio time response in order to adjust different axial forces, load torques, and rotation speeds for drive pulleys [11].

Following the calculation of power loss, the next analysis is the heat transfer simulation for the belt drive system, as heat can accumulate inside the component, leading to high temperatures and resulting in thermal failure inside the structural components of the pulley belt system. Belt-related thermal analysis has also been presented by other researchers. Merghache and Ghernaout used a numerical and experimental approach to calculate heat transfer for a synchronous belt inside a system [12]. Fortunato et al. focused on frictional heat between the belt rubber and pulley and the influence of thermal properties on heat generation [13]. Mackerle demonstrated a sliding frictional simulation of rubber-like materials using a finite element method [14]. Thermal analyses of disk-like rotating components has been the focus of other research studies. Phan and Kondyles calculated the fade cycle of a rotor and thermal coefficient by using a conjugate computational fluid dynamics (CFD) tool [15]. A dusty nanofluid thin-film flow through an inclined revolving plate at a uniform angular velocity undergoing nonlinear thermal radiation was simulated in a three-dimensional domain by Ramzan et al. [16]. The thermal analysis of a brake disk system and simulation of surface temperature plots for a rotating brake disk were performed in various studies with different focuses such as disks with fins [17] or cross-drilled holes [18], transient thermal response [19], heat conduction [20], and heat and mass transfer [21]. For instance, Wu et al. experimentally implemented the naphthalene sublimation method into a heat and mass transfer analysis of a rotating wheel [22]. Functionally graded cylinders strengthened by carbon nanotubes were used in thermoelastic analyses using a mesh-free method [23, 24]. In earlier studies, this technique was an effective approach to simulating thermoelastic behaviors during optimization of thermomechanical responses on carbon nanotube-reinforced nanocomposite sandwich plates [25, 26] and graphene-reinforced nanocomposite cylinders [27]. Thermal analysis of an entire belt drive system was also a challenging topic for researchers. Initial research by Gerbert, which focused on a simple two-pulley-one-belt drive system, presented an analytical method using Bessel functions to calculate the temperature distribution of the pulleys, which was modeled using a simplified flat disk [2]. Recent research by Wurm et al. used a new numerical approach with an updated multiple reference frame technique to calculate the surface temperature distribution of a pulley inside a continuously variable transmission system [28, 29].

One significant benefit of the thermal model (as compared with the current research) is the efficient prediction of the temperature distribution in belt drive systems without compromising accuracy. The existing method required 17 hours for thermal analysis of a belt drive under one set of operating conditions [29], which dramatically reduces design speed. In contrast, the combination of analytical and numerical methods employed in the present research guarantees that the results provided are both accurate and efficient. The thermal calculations, completed within a few seconds, meet the requirements of simulating the system under a large set of multiple operating scenarios. The thermal model provides experimentally validated temperature data within a few seconds of calculation for the design and optimization of belt drives.

In this chapter, Section 7.2 outlines the establishment of the analytical–numerical thermal model, Section 7.3 illustrates the procedures used to set up and conduct the experiment, Section 7.4 demonstrates the resultant accuracy and reliability via comparisons between the calculated and experimental temperature data, and Section 7.5 concludes this work and describes its importance in the aviation industry.

7.2 Analytical–Numerical Thermal Model

The current system uses a time-consuming CFD approach to perform thermal analyses and predict temperature distributions. The present study establishes an innovative analytical thermal model based on the thermal flow inside each pulley and the belt, as well as the heat exchanged between the two and heat generation calculations; the result is a timely prediction of the surface temperatures of the entire belt drive under each target operating condition. Calculations for the thermal behavior of the pulleys are performed using a numerical approach, due to their complex geometries and the severe turbulent flows induced by high-speed pulley rotation. Through this combination of analytical and numerical methods, the thermal model presented here can produce reliable and efficient thermal predictions for both research and industrial use.

In this section, the general overall thermal model is illustrated. Then, several thermal analyses are presented through governing equations. Finally, all equations are employed to formulate the thermal model and output temperature predictions.

7.2.1 Creation of the Analytical Thermal Model

In the beginning of every thermal analysis, the heat flux generated during operation must be determined. In pulley systems, power loss theory has been adopted to calculate heat generation because power losses are transformed into heat based on the law of the conservation of energy. The present work adopted an established power loss theory [2] to calculate power losses inside the system. According to this theory, energy losses are classified into two types: belt internal power loss W_h and contact surface power loss W_{fn}. The first loss W_h is generated at different sections of the belt and caused by the belt stretching and bending, as

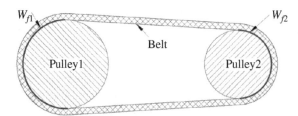

W_{f1} W_{f2}

Belt

Pulley1 Pulley2

Figure 7.2 Heat generation locations inside a two-pulley-one-belt system.

well as shear and radial compression. It can be simplified as a uniform heat source along the belt due to its high speed during operation. The second loss W_{fn} is generated at the nth belt–pulley contact surface. Figure 7.2 gives an example of the two-pulley-one-belt drive system, with W_h shown as the cross-hatched area and distributed uniformly inside the belt and W_{f1} and W_{f2}, the dark lines, indicating the belt–pulley engaged surfaces.

The thermal model demonstrated here is established based on the flow behavior and heat exchanges among the components inside the system. The system consisted of pulleys and a belt. The thermal model is divided into three subsections (Figure 7.3): internal thermal analysis of the pulleys, thermal analysis of the belt, and pulley–belt heat exchange.

First, an analysis showed that no thermal generation existed inside the pulleys, and a certain percentage of heat from W_h and W_{fn} is transmitted through the belt–pulley contact surfaces, eventually dissipating on the pulley surfaces. This work introduces the variable α_n to illustrate the percentage of frictional heat flux that flows into the nth pulley through the engaged surfaces in the belt drive system, as shown in Eq. (7.1):

$$\alpha_n W_{fn} = P_{pn} \tag{7.1}$$

where P_{pn} is the heat flux the nth pulley acquires from the frictional heat.

This amount of heat is dissipated on the pulley surfaces, and is related to the local surface temperature, exposed surface areas, thermal conductivity of the pulley material, ambient temperature T_a, operating conditions such as rotational velocities, and other potential factors. This phenomenon is represented by Eq. (7.2):

$$\alpha_n W_{fn} = f\left(\mathbf{T_{pn}}, \mathbf{D_{pn}}, \omega_{pn}, \Lambda_{pn}, C_{pn}, T_a, \cdots\right) \tag{7.2}$$

where matrix $\mathbf{T_{pn}}$ is the database storing the temperature data for the nth pulley, matrix $\mathbf{D_{pn}}$ is the pulley dimensions, ω_{pn} is the rotational speed of the nth pulley, C_{pn} is the local heat transfer coefficient, and Λ_{pn} is the material thermal conductivity of the nth pulley.

Secondly, in the belt–pulley heat exchange analysis, the flux exchanged is affected by the frictional heat flux and belt-side contact temperature T_b. This exchanged flux also influenced the heat flux flowing into the pulley. Therefore, $\alpha_n W_{fn}$ is highly related to the frictional heat flux and temperature difference at the pulley–belt surfaces, represented by the governing Eq. (7.3):

$$\alpha_n W_{fn} = g\left(\mathbf{T_{pn}}, T_b, W_{fn}\right) \tag{7.3}$$

where the engaged surface temperature at the pulley side is T_{pcn}, and the temperature data are inside matrix $\mathbf{T_{pn}}$.

Thirdly, in the internal thermal analysis of the belt, the total heat generated in the system is found to be a combination of the heat dissipated from both exposed surfaces of the belt

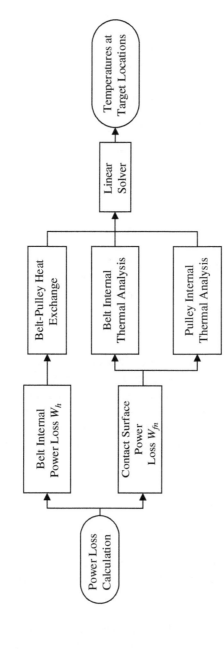

Figure 7.3 Flow chart of the calculation process for the thermal model.

and exposed surfaces of the pulleys. The heat dissipated from the belt surfaces is also highly dependent on T_a, the belt geometry, belt velocity V_b, belt thermal conductivity Λ_b, and other parameters. Therefore, the governing equation for this analysis is Eq. (7.4):

$$\sum_{n=1}^{N} (1 - \alpha_n) W_{fn} + W_h = k \left(\mathbf{T_b}, \mathbf{D_b}, V_b, \Lambda_b, C_b, T_a, \cdots \right) \tag{7.4}$$

where matrix $\mathbf{T_b}$ is the belt temperature distribution information, matrix $\mathbf{D_b}$ contains the belt's geometric dimensions, and N represents the total number of pulleys inside the transmission system.

Overall, the simple thermal model consists of Eqs. (7.2–7.4):

$$
\begin{cases}
\alpha_1 W_{f1} = f \left(\mathbf{T_{p1}}, \mathbf{D_{p1}}, \omega_{p1}, \Lambda_{p1}, C_{p1}, T_a, \cdots \right) \\
\quad \vdots \\
\alpha_n W_{fn} = f \left(\mathbf{T_{pn}}, \mathbf{D_{pn}}, \omega_{pn}, \Lambda_{pn}, C_{pn}, T_a, \cdots \right) \\
\alpha_1 W_{f1} = g \left(\mathbf{T_{p1}}, T_b, P_{f1} \right) \\
\quad \vdots \\
\alpha_n W_{fn} = g \left(\mathbf{T_{pn}}, T_b, P_{fn} \right) \\
\sum_{n=1}^{N} (1 - \alpha_n) W_{fn} + W_h = k \left(\mathbf{T_b}, \mathbf{D_b}, V_b, \Lambda_b, C_b, T_a, \cdots \right)
\end{cases}
\tag{7.5}
$$

In this model, functions $f(\mathbf{T_{pn}},...)$, $g(\mathbf{T_{pn}}, ...)$, and $k(\mathbf{T_b},...)$ must be defined to output the temperature distribution. The following subsections will illustrate the establishment of each function using analytical and numerical approaches.

7.2.2 Belt Thermal Analysis

The cross-section of the belt is assumed to be rectangular in shape, making it possible for the heat transfer analytical method to be applied. In this study, it is reasonable to assume that the belt had a uniform temperature distribution because the ratio of thickness to length was above 10, and the thermal conductivity of the rubber is very low (0.8 W/(m·K)). Consequently, the entire belt had a single temperature, represented as T_b, which replaced the matrix $\mathbf{T_b}$. This simplification also streamlined the establishment of a thermal governing equation for the belt [30], as shown in Eq. (7.6):

$$\left(T_b - T_a \right) S_b C_b = W_h + \sum_{n=1}^{N} W_{fn} \left(1 - \alpha_n \right) \tag{7.6}$$

where the surface area exposed to the surrounding area is S_b, and the local heat transfer coefficient is C_b.

C_b is related to the velocity of the local air and thus is measured experimentally on a small wind tunnel device. The polynomial Eq. (7.7) is used to describe the resultant relationship:

$$C_b = p_1 V_b^4 + p_2 V_b^3 + p_3 V_b^2 + p_4 V_b + p_5 \tag{7.7}$$

where p_1, p_2, p_3, p_4, and p_5 represent the five coefficients acquired by curve-fitting with their resulting values: $p_1 = -6.9 \times 10^{-5}$, $p_2 = 7.1 \times 10^{-3}$, $p_3 = 2.2 \times 10^{-1}$, $p_4 = 4.3$, and $p_5 = 1.7$. The thermal model calculated C_b based on V_b and then substituted it into Eq. (7.6) for the thermal calculation.

7.2.3 Heat Exchange at the Pulley–Belt Contact Surfaces

The frictional heat generation and pulley–belt heat exchange coexisted at the surfaces engaged between the belt and pulley. There is no heat exchange when T_{pcn} is identical to T_b and $\alpha_n = 0.5$. However, heat flowed into the pulley from the belt side and $\alpha_n > 0.5$ when $T_b > T_{pcn}$, and vice versa. These phenomena can be described by Eq. (7.8):

$$P_{exn} = (\alpha_n - 0.5)\, W_{fn} \tag{7.8}$$

where the exchanged heat flux per second from the belt to the nth pulley is P_{exn} and $P_{exn} < 0$ means there is heat flow from the pulley to the belt. Also, P_{exn} is dependent on the temperature difference between the pulley and belt, as shown in Eq. (7.9):

$$P_{exn} = S_{exn} C_{exn} (T_b - T_{pcn}) \tag{7.9}$$

Combining Eqs. (7.8, 7.9) defines the function $g(\mathbf{T_b}, T_{pcn}, W_{fn})$ in Eq. (7.10):

$$S_{exn} C_{exn} (T_b - T_{pcn}) = (\alpha_n - 0.5)\, W_{fn} \tag{7.10}$$

7.2.4 Pulley Internal Thermal Analysis

The following analysis formulates the relationship between the functions $f(\mathbf{T_{pn}},...)$ and Φ_{pn}. The optimal approach in this analysis is the numerical rather than analytical method because it is not limited to restrictions when solving for complex pulley structures and dimensions. Moreover, the numerical method is currently the only option for modeling turbulence flows induced by high-speed rotations, a necessary step in calculating the local airspeed and accurate computation of the heat dissipation flux (attributable to the local airspeed's influence on the local heat dissipation coefficients). The numerical software used in this work is ANSYS® 19.1.

However, the numerical method could not efficiently provide the resultant temperature distributions because of the high computational cost. Thus, this work investigates the thermal flow inside the pulley by examining and determining the factors impacting P_{pn}. First, a mathematical algorithm is formulated to illustrate the relationship between P_{pn} and potential parameters. Then, these parameters are divided into different groups based on their changes. Finally, numerical simulations are adopted to create an accurate relationship by establishing a resultant database.

7.2.4.1 Mathematical Algorithm
The mathematical algorithm cannot be applied to a geometrically complex pulley; thus, some insignificant geometric features had to be simplified, as shown in Figure 7.4. After suppressing the grooved surfaces and fillets, this work formulates the algorithm on a round I-structure with fins. The purpose of this section is to illustrate the creation of the algorithm based only on the primary I-structure and then expand it to a calculation of the fin structure.

To analyze the thermal flow within the pulley, the complex I-shape structure is divided into three zones (Figure 7.5) and reduced to three rectangular cross-sections. In each section, the temperature's governing is established using the Bessel function in Eq. (7.11).

$$\frac{d^2 T}{dr^2} + \frac{1}{r}\frac{dT}{dr} - \frac{C_p}{B_i \Lambda_p}(T - T_a) = 0 \tag{7.11}$$

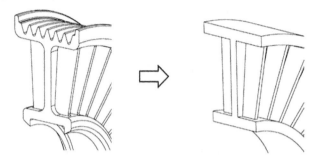

Figure 7.4 Simplification of the pulley structure. Source: [33]

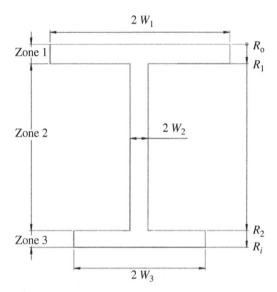

Figure 7.5 Three zones of the I-structured pulley.

where $T(r)$ represents the temperature curve along the pulley's radial direction.

However, the pulley's fins influence heat dissipation in Zone 2. An infinitesimal element dr serves here as an example. The surfaces of the fins amplified the amount of heat dissipation, increasing the heat flow on the fins' connected areas (S_f in Figure 7.6) compared to areas exposed to the environment (S_n in Figure 7.6). In this study, the ratio of the heat flux dissipated between S_f and S_n is represented by a new coefficient λ_{fin}. When the fins play no role in heat dissipation, the dissipation rates on S_f are the same as on S_n, leading to a dissipation rate on the two sides of the infinitesimal element of: $4\pi r \cdot dr \cdot C_p$. When the fins have some influence on the heat distributed on the pulley's surface, the dissipation rate on the two sides of this infinitesimal element became $2 \cdot [N_{\text{fin}} \cdot W_{\text{fin}} \cdot dr \cdot \lambda_{\text{fin}} \cdot C_p + (2\pi r \cdot dr - N_{\text{fin}} \cdot W_{\text{fin}} \cdot dr) \cdot h_p]$. The amplified ratio is $[1 + N_{\text{fin}} \cdot W_{\text{fin}} \cdot (\lambda_{\text{fin}} - 1)/2\pi r]$. Therefore, it is reasonable to assume that the local h_p on the two side surfaces of Zone 2 also increased according to this ratio. Hence, Eq. (7.11) is:

$$\frac{d^2 T}{dr^2} + \frac{1}{r}\frac{dT}{dr} - \frac{[2\pi r + N_{\text{fin}} B_{\text{fin}}(\lambda_{\text{fin}} - 1)] C_p}{2\pi r B_i \Lambda_p}(T - T_a) = 0 \tag{7.12}$$

The dimensions of the fins determine the value of λ_{fin}. The fin section in the example of a rounded infinitesimal element could also treated as a rectangle beam whose temperature

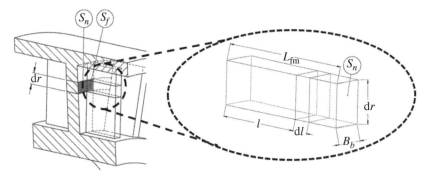

Figure 7.6 Fin geometry.

distribution equation $G(l)$ could then be established based on the heat dissipated on the exposed surfaces of the beam and heat flux, with S_{fin} being equivalent [31]:

$$\frac{\partial^2 G}{\partial l^2} = -\frac{2C_p\left(G - T_a\right)}{\Lambda_p B_{\text{fin}}} \tag{7.13}$$

The value of $G(0)$, which shares the boundaries and local temperatures with the infinitesimal element at radius r, represents the temperature at S_f and $T(r)$. The total heat dissipated from this fin is provided by the temperature integration of all exposed surfaces and the far-end side (S_e in Figure 7.6).

$$P_{\text{fin}} = \left\{ 2 \int_0^{L_{\text{fin}}} G(l)\, dl + B_{\text{fin}} \cdot \left[T(r) - T_a\right] \right\} H_{\text{fin}} C_p \tag{7.14}$$

According to the definition of λ_{fin},

$$\lambda_{\text{fin}} = \frac{P_{\text{fin}}}{\left[T(r) - T_a\right] H_{\text{fin}} B_{\text{fin}} C_p} \tag{7.15}$$

The solution, which is λ_{fin} for this infinitesimal element, is developed as:

$$\begin{cases} \lambda_{\text{fin}} = \dfrac{2\sinh(mL_{\text{fin}}) + mB_{\text{fin}}}{mB_{\text{fin}}\cosh(mL_{\text{fin}})} \\[2mm] m = \sqrt{C_p/\Lambda_p B_{\text{fin}}} \end{cases} \tag{7.16}$$

The above equation demonstrates that λ_{fin} is a function of the fin geometries (L_{fin}, W_{fin}), Λ_p and C_p, and not the radius r.

Equation (7.12) could then be transformed into one kind of Kummer's equation. Its analytical solution is provided in [32], and is also the definition of $T(r)$.

$$\begin{cases} T(r) = U_1(r) \cdot K_1 + U_2(r) \cdot K_2 + T_a \\[4mm] U_1(r) = e^{-\frac{\sqrt{C_p}r}{\sqrt{B_i\Lambda_p}}} \text{KummerM}\left[\dfrac{\frac{\sqrt{C_p}N_{\text{fin}}B_{\text{fin}}(\lambda_{\text{fin}}-1)}{2} + \pi\sqrt{\Lambda_p B_i}}{2\pi\sqrt{\Lambda_p B_i}}, 1, \dfrac{2\sqrt{C_p}r}{\sqrt{B_i\Lambda_p}}\right] \\[6mm] U_2(r) = e^{-\frac{\sqrt{C_p}r}{\sqrt{B_i\Lambda_p}}} \text{KummerU}\left[\dfrac{\frac{\sqrt{C_p}N_{\text{fin}}B_{\text{fin}}(\lambda_{\text{fin}}-1)}{2} + \pi\sqrt{\Lambda_p B_i}}{2\pi\sqrt{\Lambda_p B_i}}, 1, \dfrac{2\sqrt{C_p}r}{\sqrt{B_i\Lambda_p}}\right] \end{cases} \tag{7.17}$$

The value of the constant coefficients K_1 and K_2 are calculated through the boundary conditions. The inner temperature of this zone $T(R_{li})$ is a non-infinity value, making K_2

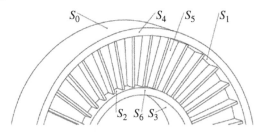

Figure 7.7 Pulley outer surfaces. Source: [33]

equal to 0 in order to discard the infinity value of function $U_2(R_{li})$ in Eq. (7.17). Then K_1 could be calculated as:

$$
\begin{cases}
r = R_{lo} \Rightarrow K_1 = \frac{T(R_{lo}) - T_a}{F_1(R_{lo})} \\
r = R_{li} \Rightarrow T\left(R_{li}\right) = \text{limited}; K_2 = 0
\end{cases}
\tag{7.18}
$$

The temperature distribution for one zone is then:

$$
T(r) = \frac{U_1(r)}{U_1\left(R_{lo}\right)} \left[T\left(R_{lo}\right) - T_a\right] + T_a
\tag{7.19}
$$

The pulley's temperature distribution could then be calculated using the above equation, zone by zone. This equation demonstrates that the value of $T(R_{lo})$ determined function $T(r)$ in one zone. At the same time, the shared boundary of the two zones has the same temperature, which meant $T(R_{lo})$ in one zone is equivalent to $T(R_{li})$ in the previous zone. Therefore, the pulley's temperature distribution is determined by T_{pcn}, which is the temperature at the outer radius and the temperature within the belt. So long as this value is known, Eq. (7.19) could be used to calculate the values of $T(R_{lo})$ for Zones 2 and 3, as well as the values of $T(R_{li})$ for Zones 1 and 3, thus providing the whole temperature distribution $\mathbf{T_{pn}}$.

After acquiring the temperature distribution function, the heat dissipation rate function could be calculated. For the heat dissipation rate at the side surfaces (S_4, S_5, and S_6 in Figure 7.7), P_{ps} is:

$$
P_{ps} = Q\left(R_o, R_1\right) + Q\left(R_1, R_2\right) + Q\left(R_2, R_i\right)
\tag{7.20}
$$

where the pulley's outer and inner radii are R_o and R_i, respectively, and R_1 and R_2 are the inner radii of Zones 1 and 2, respectively (Figure 7.5). Thus, the dissipation rate at the side surfaces of each zone could be described as:

$$
Q\left(R_{lo}, R_{li}\right) = 2\int_{R_{li}}^{R_{lo}} C_p \left[2\pi r + N_{\text{fin}} B_{\text{fin}} \left(\lambda_{\text{fin}} - 1\right)\right] \left[T(r) - T_a\right] dr
\tag{7.21}
$$

Furthermore, the heat dissipation rate for the rest of the surfaces P_{pb} is:

$$
P_{pb} = (1 - \varphi/2\pi) S_0 C_{pn} \left[T\left(R_o\right) - T_a\right] + S_1 C_{pn} \left[T\left(R_1\right) - T_a\right] + \\
S_2 C_{pn} \left[T\left(R_2\right) - T_a\right] + S_3 C_s \left[T\left(R_i\right) - T_a\right]
\tag{7.22}
$$

where surface areas S_0, S_1, S_2, and S_3 are as defined in Figure 7.7; the wrap angle, which is the amount of belt in contact with the pulley, is φ; and the shaft–pulley heat transfer coefficient is C_s. The heat dissipation rate on all pulley surfaces can be written as:

$$
P_{pn} = P_{pb} + P_{ps}
\tag{7.23}
$$

The above algorithm outlines the procedure for calculating $\mathbf{T_{pn}}$ and provides the definition of the function $f(\mathbf{T_{pn,...}})$ used to acquire P_{pn}. There are more than 10 input parameters for this calculation. To simplify the calculation, these parameters are categorized into two sets. One set is pulley-related parameters, which included R_i, R_2, R_1, R_o, N_{fin}, L_{fin}, W_{fin}, H_{fin}, S_0, S_1, S_2, S_3, S_{fc}, S_4, S_5, and S_6, as well as the material-related parameter Λ_{pn}. These values are geometry- and material-related parameters that remained constant during the analysis. The other set of parameters included C_{pn}, ω_{pn}, and T_{pcn}, which are variable parameters related to the current operating condition. C_{pn} is an experimental function of ω_{pn}. Therefore, for a specific pulley, P_{pn} could be determined by only two variable parameters: ω_{pn} and T_{pcn}.

The above algorithm also indicates that P_{pn} has a linear relationship with T_{pcn} when ω_{pn} remains a constant value. This is because the Reynolds number and turbulence severity do not change at the constant spinning speed of the pulley. P_{pn} has a nonlinear relationship with ω_{pn} when T_{pcn} remains the constant value. The Reynolds number and severity of the turbulence flow do not have an exponential relationship with rotation speed. Thus, this work introduces η_{pn}, denoted as the nth pulley's thermal coefficient to characterize the ratio of P_{pn} to T_{pcn}. The value of this coefficient represents the extent of the heat dissipation that a pulley can achieve. There is also a function defined as $E_n(\omega_{pn})$ which is used to describe how the value of η_{pn} changes at different rotational speeds. Therefore, a new equation is developed as:

$$\alpha_n W_{fn} = \left(T_{pcn} - T_a\right) E_n\left(\omega_{pn}\right) \tag{7.24}$$

The numerical approach, which can accurately calculate the turbulence flow and local air velocity, is adopted to calculate the value of η_{pn} when ω_{pn} increases from 2000 to 6000 RPM at intervals of 1000 RPM. $E_n(\omega_{pn})$ is then calculated by performing a curve-fitting by using a fourth-order polynomial on these values. The temperature distribution from the numerical approach under baseline conditions is used as a database for outputting the thermal distributions at the target operating condition, as illustrated in the following section.

7.2.4.2 Numerical Method
To calculate η_{pn} and the function $E_n(\omega_{pn})$, the numerical approach must be different from the mathematical algorithm. This method involves geometric modeling, meshing techniques, boundary condition setup, and optimized CFD settings.

7.2.4.2.1 Modeling Structure
The numerical study used solid regions for components and a fluid region for the air. Figure 7.8 displays the geometries of both regions. The solid region contains geometries of the shaft and pulley to perform thermal conduction within these parts. To simulate the airflow, a round fluid region encloses the solid parts. Its diameter is much larger than the pulley diameter to provide enough simulation domain to accurately model the flow and vortices around the pulley. The belt is neglected because it is a thin structure and had limited influence on airflow. A special treatment (Section 7.2.4.2.5) is used to simulate the portion of the frictional heat flux entering the pulley areas.

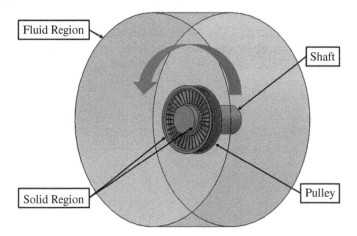

Fluid Region

Shaft

Solid Region

Pulley

Figure 7.8 Regions and geometries of the numerical simulation. Source: [33]

7.2.4.2.2 Modeling Motions

This study adopted a single reference frame (SRF) to simulate the pulley and shaft rotating against the surrounding air. This technique is well suited for this static-state analysis because the shapes of both the fluid and solid domains remained unchanged. Unlike real-world conditions where the solid parts rotate against the fluid media, this approach has the fluid media moving against the stationary solid part. Although relative speeds are the same in both situations, the latter approach dramatically simplified the numerical problem from the fluid–structure interaction to a static thermal analysis. The fluid domain could then be used to simulate the airflow adjacent to the pulley, while the solid domain is employed to simulate the heat conduction within the pulley.

There are alternatives to an SRF that are commonly employed. One option, the mixing plane method, focuses on a single radial profile using a circumferential averaging technique, which in turn extends to all radial profiles at the interface. It is often used to simulate the flows triggered by a rotor rotating against a stator inside a flow processor. Another technique, the sliding mesh method, considers the unsteady flow due to the rotating components moving against the stationary parts. A third option, the dynamic mesh method, adjusts the mesh based on the moving boundaries and objects. These transient and computationally expensive approaches are not suitable for the steady-state condition in this study. Therefore, the selection of an SRF is preferable for conducting the numerical simulations presented in this chapter.

7.2.4.2.3 Meshing Procedure

This study uses the same meshing plot method for both the solid and fluid regions. First, the triangular surface mesh is generated and converted to polyhedrons for three-dimensional computation. In this analysis, the minimum edge length is set to 0.2 mm and the growth rate to 1.1. The belt–pulley grooved surfaces contacted are critical and optimized with a relatively dense mesh, while the rest of the area has a coarser mesh, as shown in Figure 7.9. Moreover, this simulation uses conformal mesh to bond the solid and fluid domains

Figure 7.9 Polyhedral mesh used in the numerical analysis.

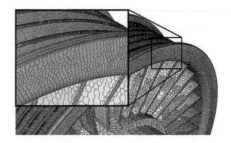

together to facilitate the convergence of the numerical approach and enhance the computational stability. This is due to these meshes sharing boundary nodes and element edges.

7.2.4.2.4 Selection of the Heat Transfer Model

The challenge for setting up the heat transfer model involves the selection of turbulence models and solution method. This selection determines whether this numerical approach could simulate the pulley rotation-induced airflow and provide local air velocities to accurately acquire the local heat dissipation rate of the exposed surfaces. In this study, the Menter Shear Stress Transport (SST) k–ω model, a submodel of the Reynolds Averaged Navier–Stokes (RANS), is selected for the airflow simulation because of its acceptable near-wall treatment, reasonable computational time, and high stability. Moreover, the incompressible ideal gas equation is selected for application. Air is a fluid medium, and 140 °C is the maximum temperature FRP can reach before failure. The air velocity is subsonic at the highest rotational speed in the system, corresponding to a 108 mm pulley rotating at 7000 RPM. The density of air could be considered as a constant. Hence, the finite volume equations [31] shown below solve the governing equations in integral form:

$$\frac{\partial}{\partial t} \int_V^j M dV + \oint [J - O] \cdot da = \int_V^j H dV \tag{7.25}$$

with:

$$\mathbf{M} = \begin{bmatrix} \rho \\ \rho v \\ \rho Z \end{bmatrix}; \mathbf{J} = \begin{bmatrix} \rho \left(v - v_g \right) \\ \rho \left(v - v_g \right) \otimes v + \rho \mathbf{I} \\ \rho \left(v - v_g \right) \mathbf{H} + p v_g \end{bmatrix}$$

$$\mathbf{O} = \begin{bmatrix} 0 \\ Q \\ Qv + \dot{q}'' \end{bmatrix}; \mathbf{H} = \begin{bmatrix} S_u \\ g \\ S_u \end{bmatrix}$$

where the vector of conserved mass, momentum, and energy is **M**. The vector of inviscid terms, viscous terms, and body forces are **J**, **O**, and **H**, respectively. The p variable represents the pressure, and the heat flux is $\dot{q}''$. The ρ and v variables are the density and velocity vector of the air, respectively. The vector of velocity for the grid is v_g, and Z is the overall energy per unit mass. The SST k–ω model consisted of specific dissipation rate equations and turbulence kinetic energy, as shown by [31]:

$$\frac{\partial}{\partial t} (\rho k) + \frac{\partial}{\partial \chi} (\rho k v) = \frac{\partial}{\partial \chi} \left(\Gamma_k \frac{\partial k}{\partial \chi} \right) + G_k - Y_k \tag{7.26}$$

$$\frac{\partial}{\partial t} (\rho \omega) + \frac{\partial}{\partial \chi} (\rho \omega v) = \frac{\partial y}{\partial \chi} \left(\Gamma_\omega \frac{\partial \omega}{\partial \chi} \right) + G_\omega - Y_\omega + D_\omega \tag{7.27}$$

where the effective diffusivity and kinetic energy of k and ω are Γ_k and Γ_ω and G_k and G_ω, respectively, and Y_k and Y_ω are the turbulence-induced dissipation of k and ω, respectively. The extra term in Eq. (7.27), D_ω, is a blend function to transform the equation either to the k–ω model at locations near the wall or the k–ω model in the far field. This ensures that in the present research, the equation has robust near-wall treatment and a free stream away from the wall. The default constants are used.

The second-order upwind option is chosen when setting the spatial discretization. This activates the pseudo-transient function, guaranteeing its rapid convergence and better stabilizing these cases. Another setting is the initial temperature, which is set to the operating environmental temperature; the thermal conductivities of the materials of the belt and pulleys are acquired from an experiment based on ISO Standard 22007-2.

7.2.4.2.5 Input/Output Setting

The simulation of the flux P_{pn} flowing into the pulley during the numerical study is outlined here. The grooved surfaces at the pulley's outer radius have a continuously changing flow during rotation. When the surface engaged with the belt, the pulley obtained a certain percentage of generated frictional heat, and this thermal flow flux to the pulley is positive. When it is disengaged with the belt after rotating a set amount, this surface dissipated the heat and the thermal flow flux to the pulley is negative. There is no value or direction change in the settings of the boundary conditions during numerical simulation. Additionally, this change is at a high frequency, due to the pulley's high-speed rotation. Consequently, a special treatment is used and displayed in Figure 7.10 [33]. The flat and small sections adjacent to the engaged surfaces are used to generate frictional heat in the numerical simulation. These sections barely influenced the temperature distribution because they are small and close to the surface. Figure 7.10 also shows areas exposed to the environment that only dissipated heat in ways similar to the other pulley surfaces. The boundary of the fluid domain is considered an opening pressure outlet and set to zero-gauge pressure. The backflow turbulence intensity to viscosity ratio is set with default values of 5 and 10%, respectively. All boundaries are set to the nearby operating temperature.

The output of this simulation includes the following information at the baseline conditions: the temperature distribution T'_{pn}, average temperature of the contact surfaces at the pulley sides T'_{pcn}, and total heat flux dissipated from the surfaces of the pulley P'_{pn}. The current numerical simulation establishes the curve of $E_n(\omega_{pn})$, and the conditions can act as a baseline for future predicted temperatures under various targeted operating conditions.

7.2.4.2.6 Parametric Setting

This section illustrates the parametric numerical simulations and their results on being integrated into the thermal model. The mathematical model has already demonstrated how changes in T_{pcn} and ω_{pn} influenced the value of η_{pn} for each pulley and Eq. (7.24) in the thermal model calculation. Figure 7.11 compares the curve of η_{pn} from both numerical and mathematical calculations for a 108 mm pulley rotating at a speed of 6000 RPM with contact temperatures varying from 40 to 80 °C, under a constant temperature T_{pcn} of 50 °C, with the rotating speed varying from 2000 to 6000 RPM at 1000 RPM intervals. Both methods prove that η_{pn} is constant when T_{pcn} is changed but it increased with the increase of ω_{pn}. In Figure 7.11, the values of η_{pn} obtained from the mathematical method are lower than those from the numerical algorithm. This is because the mathematical algorithm simplified the grooved features into a flat surface, as shown in Figure 7.4, decreasing the surface area

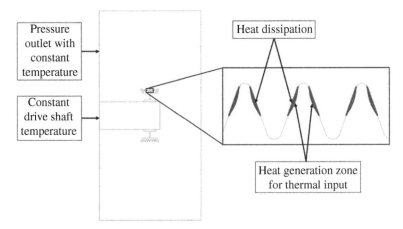

Figure 7.10 Special treatment of heat generation and dissipation at engaged surfaces. Source: [33]. Licensed under CC BY 4.0.

of S_0 denoted in Figure 7.7 and its surface dissipation P_{pn} in Eq. (7.22). This proves that the values of η_{pn} from the numerical method are more accurate, which is the reason the numerical method is selected for this research.

This study uses the numerical simulation to calculate η_{pn} at each baseline condition at speed increments of 1000 RPM to provide datapoints for the curve-fitting equations of $E_n(\omega_{pn})$. Moreover, $\mathbf{T}'_{pn}$ and Φ'_{pn} are useful for the pulley's thermal distribution calculations. Consequently, a group of data $\{E_n(\omega_{pn}), \mathbf{T}'_{pn}, P'_{pn}\}$ for each pulley is stored in a database. The interval between the two speeds is called a speed zone.

7.2.5 Overall Structure

The thermal model designed for belt drive systems is formulated by a combination of Eqs. (7.6, 7.10, 7.24):

$$
\begin{cases}
\displaystyle\sum_{n=1}^{N} \left(1 - \alpha_n\right) P_{fn} + P_h = S_b C_b \left(T_b - T_a\right) \\
S_{ex1} C_{ex1} \left(T_b - T_{pc1}\right) = \left(\alpha_1 - 0.5\right) W_{f1} \\
\quad\vdots \\
S_{exN} C_{exN} \left(T_b - T_{pcN}\right) = \left(\alpha_N - 0.5\right) W_{fN} \\
\alpha_1 W_{f1} = E_1\left(\omega_{p1}\right)\left(T_{pc1} - T_a\right) \\
\quad\vdots \\
\alpha_N W_{fN} = E_N\left(\omega_{pN}\right)\left(T_{pcN} - T_a\right)
\end{cases}
\tag{7.28}
$$

The lower–upper decomposition serves as an efficient method of solving the above system of equations and outputting the contact temperatures on both the pulley and belt sides. The total calculation time is generally instant on a computer with an AMD 3950X CPU.

To obtain the temperature profiles for the entire system, additional calculations are required. The belt temperature is equivalent to the contact temperature on the belt side, based on the assumption defined at the beginning of this chapter. The pulley temperature distribution, matrix $\mathbf{T_{pn}}$, is calculated by adjusting the baseline numerical results. The model identified ω_{pn} and determined within which speed zone it is located, and then

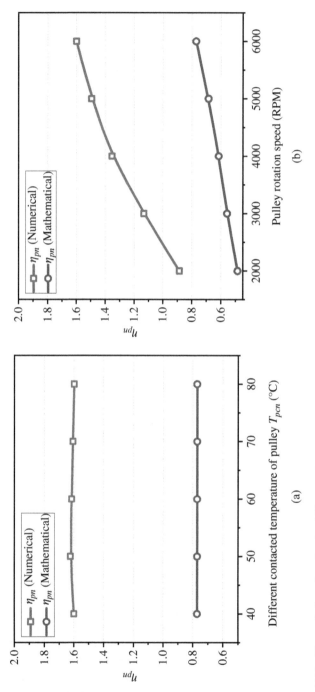

Figure 7.11 The variation of η_{pn} for the 108 mm pulley versus (a) T_{pcn} when ω_{pn} = 6000 RPM and (b) ω_{pn} when T_{pcn} = 60 °C, as obtained from numerical simulations and analytical model.

imported the two sets of $\{\mathbf{T}'_{pn}, P'_{pn}\}$ of the investigated pulley at the upper and lower speeds of the speed zone. The model then calculated P''_{pn} and $\mathbf{T}''_{pn}$ at the planned speed ω_{pn} via a linear interpolation. The temperature distribution of the nth pulley is calculated as:

$$\mathbf{T}_{pn} = \alpha_n W_{fn}/P''_{pn} \left(\mathbf{T}''_{pn} - T_a\right) + T_a \tag{7.29}$$

7.3 Experimental Setup

This study also validated the accuracy of this model under various operating conditions for both research and industrial use. A rigorous experimental arrangement is created prior to certifying the thermal model under a number of configurations. The setup of these experiments involves several tasks, including belt drive layout, several load cases (such as speed changes), and the design of related equipment. Later, experiments are performed to measure the temperature data, which are then compared with the results calculated under identical operating conditions.

7.3.1 Operating Conditions

Consideration of a range of testing conditions is critical for this thermal model's validation. The pulleys used in this application are considerably different in terms of materials and radii. In these experiments, two different materials and two different radii of pulleys are selected. Speed changes are also considered, and three speed test points are selected to consider variations in the results from the idle and highest RPM situations. Table 7.1 summarizes the configuration for parameter variations in a number of operating conditions. Both steel and composite pulleys are fabricated. Transmitted speed is controlled in a dynamometer by altering the revolution of the DR pulley in the belt drive system.

7.3.2 Belt Drive Layout

The design of the belt drive layout is also essential to the experimental setup. A belt drive system with a minimum of one DR and DN pulley is sufficient for this study, plus three auxiliary pulleys used to measure and stabilize the high-speed belt drive system, which could reach as high as 6000 RPM. The radius of the DR steel pulley is 148 mm, and that of the DN FRP pulley is 108 mm. Table 7.2 provides the locations of pulleys within the belt drive systems that are considered for this experiment.

7.3.3 Equipment Setup

The belt drive system, together with the measurement devices and data acquisition system, is installed on a modified engine dynamometer. The DR pulley, denoted as Pulley 1, is located at the bottom left. The DN pulley, designated as Pulley 4, is at the top middle of the layout. The hub load pulley, referred to as Pulley 2, is located at the bottom right, measuring the belt tension from the hub load sensor beneath the pulley. Pulley 3 is located at the far right, measuring the belt speed through a speedometer underneath it. Pulley 5, located at

Table 7.1 Parameters investigated to validate the thermal model.

Parameter	Symbol	Variation
Radius of the pulley (mm)	R	108/148
Thermal conductivity of the material (W/(m·K))	Λ_p	55 (steel)/0.3 (FRP)
Transmitted speed of DN pulley (RPM)	ω_{p4}	2000/4000/6000

Table 7.2 Locations of the tested belt drive pulleys.

Pulley ID	X (mm)	Y (mm)	Radius (mm)
1	0	0	148
2	240	−29	65
3	311	255	62
4	205	185	108
5	96	113	65

the upper left, is used to reduce the belt vibration and maintain belt tension by connecting a tensioner. The dynamometer is modified by equipping it with an insulated chamber with a constant temperature system using two thermocouples inside the insulated chamber to read the temperature and an air blower on top to increase the temperature, if needed. When two readings are quite different, the case fan is introduced at the bottom of the chamber to keep the temperature uniformly distributed. Several thermal measurement devices are also equipped at the target locations. There are five infrared cameras mounted to acquire the temperature data at preset locations. A thermal camera is used to record the surface temperature of the whole DN pulley. Figure 7.12 presents the installed belt drive system inside the insulated chamber before the experiment [33]. The experiments are performed three times under all operating conditions. All collected temperature data are recorded and averaged by an ancillary data acquisition system for digital processing.

Simultaneously, the thermal model inputs the same thermal properties and pulley geometries (Λ_p, h_p, ...) to predict the temperature distribution according to the abovementioned operating conditions and setup.

7.4 Results and Discussion

This section investigates the temperature prediction performances of the model under various material types, pulley radii, and rotational velocities, as described in Table 7.1. The reliability of λ_{pn} and the temperature distribution under baseline conditions is validated to prove that the precalculated database provided by the numerical simulations is accurate enough to be imported and utilized by the thermal model. Then, a comparison of the

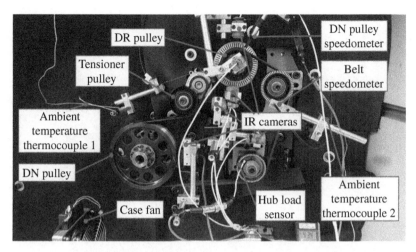

Figure 7.12 Installation of the belt drive before the experiment. Source: [33].

experimental results is made with the thermal model in terms of the temperature at various locations across the belt drive system. Finally, the temperature profile for each pulley is analyzed via temperature comparisons at different locations along the radius of the pulley.

7.4.1 Verification of the Belt's Uniform Temperature

The assumption of the belt's uniform temperature distribution is a fundamental prerequisite for the thermal model. It is necessary to prove this assumption via the experiment before comparing additional results. The experiments shown in Figure 7.13a,b demonstrate that the front and back sides of the belt had nearly identical temperatures. A temperature profile from the numerical approach performed for a different application [33] also verifies the assumption of belt uniformed temperature distribution, as shown in Figure 7.13c. Consequently, this assumption is practical for this thermal model.

7.4.2 Verification of Curve of $E_n(\omega_{pn})$

The values provided by $E_n(\omega_{pn})$ are critical to achieving the subsequent temperature results from the thermal model. These are validated by comparing the measured and numerical data of the heat dissipation flux rate for each pulley under the condition of the same temperatures at the belt–pulley engaged surfaces. It is impossible to measure such a heat dissipation flux rate with the pully at such a high RPM. Therefore, the surface temperatures of the pulley are compared between the calculated and measured results because the heat dissipation flux rate is strongly linked to temperature distribution through the local heat dissipation coefficient. Figure 7.14 shows that the temperature plots of the pulleys from the numerical method and experiments are nearly identical when the pulley–side temperature is set to the same value.

The above comparison requires an accurate local heat dissipation coefficient C_p. This section describes the airflow around the pulley to certify that the predicted value of C_p is reasonable; this value is closely related to local air speed and an accurate airflow simulation

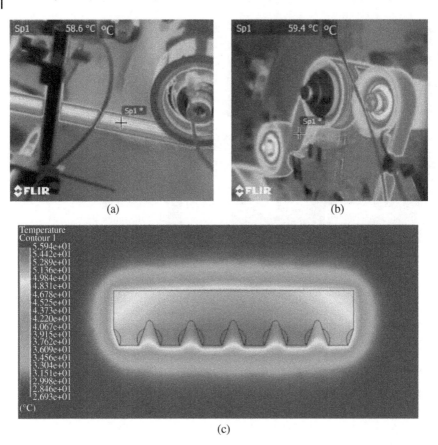

Figure 7.13 Uniform thermal distribution inside the belt. Temperatures on (a) the front surfaces of the belt and (b) the back surfaces of the belt. (c) Temperature plot calculated from the simple numerical approach. Source: (a,b) [33]. (c) [33]. Licensed under CC BY 4.0.

is a necessity provided by the numerical approach. Figure 7.15 presents the numerical airflows adjacent to the 108 mm pulley at a rotation speed of 6000 RPM, illustrating that the SST k–ω model is well suited for this application, combining both advantages of the k–ω and the k–ω models by using a blending function. The merit of the k–ω model is the accurate flow simulation in the viscous sublayer near the wall, and the superiority of the k–ω model is the accurate free flow predictions in regions away from the wall. Additionally, the time-efficient turbulence model used in this simulation makes the calculation time for each baseline condition less than three hours on a workstation with an AMD 3950 CPU, making the total time to create the precalculated database acceptable for researchers. Therefore, the turbulence model adopted here can provide reliable airflow simulations, making $E_n(\omega_{pn})$ eligible for subsequent thermal model calculations.

Several factors influence the curves of $E_n(\omega_{pn})$, providing a general design guideline for increasing the value of $E_n(\omega_{pn})$ to maximize the ability of the heat dissipation rate of the pulley. Figure 7.16 gives the values of $E_n(\omega_{pn})$ for three different pulleys, from 2000 to 6000 RPM, calculated from the numerical method. The fact that the 148 mm FRP pulley

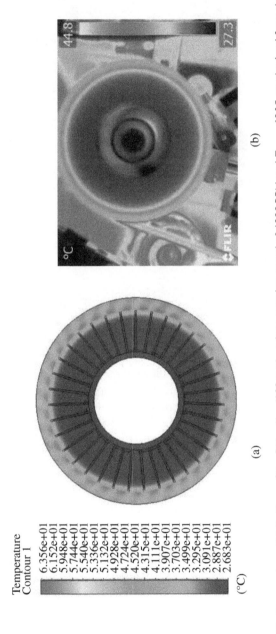

Temperature
Contour 1

6.356e+01
6.152e+01
5.948e+01
5.744e+01
5.540e+01
5.336e+01
5.132e+01
4.928e+01
4.724e+01
4.520e+01
4.315e+01
4.111e+01
3.907e+01
3.703e+01
3.499e+01
3.295e+01
3.091e+01
2.887e+01
2.683e+01
(°C)

(a)

(b)

Figure 7.14 Temperature distributions of the DN pulley (108 mm) under a rotation speed of 4000 RPM and T_{pcn} = 60 °C, as obtained from the (a) numerical and (b) experimental results.

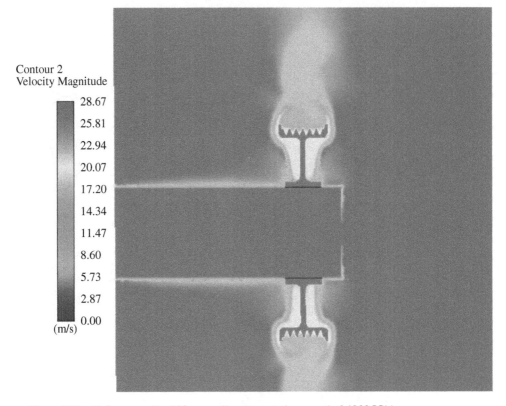

Contour 2
Velocity Magnitude

28.67	
25.81	
22.94	
20.07	
17.20	
14.34	
11.47	
8.60	
5.73	
2.87	
(m/s)	0.00

Figure 7.15 Airflow near the 108 mm pulley at a rotation speed of 6000 RPM.

having a higher η_{pn} than the 108 mm FRP pulley indicates that the diameter of the pulley increased the value of $E_n(\omega_{pn})$. This is because large dissipation areas with sizeable diameters provided extra heat dissipation area, increasing the value of η_{pn}. Λ_p also exerts a similar influence because materials with high thermal conductivity are prone to conduct heat from inside to the outer surface for quick heat dissipation, as shown by the fact that the 148 mm steel pulley has a higher η_{pn} than the 108 mm FRP pulley. Therefore, increasing the pulley radius or adopting the high Λ_p can efficiently reduce the surface temperatures of pulleys and prolong the life expectancy.

7.4.3 Model Verification

Validation is established by comparing the experimental results for the temperatures on both sides of the belt–pulley engaged surface with those obtained from the calculation of the thermal model. The highest temperatures are observed at the component contact surfaces, making it critical to predicting their structural failure. Figure 7.17 displays the comparison between the analytical and experimental results for the contact temperatures of the 108 mm pulley, 148 mm pulley, and belt inside the belt drive system at rotation speeds for the DN pulley ranging from 2000 to 6000 RPM. The thermal model predicts an upward trend in temperature with increasing DN pulley rotation speed, which matches the trend observed in the

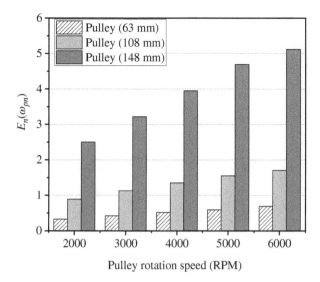

Figure 7.16 Values of $E_n(\omega_{pn})$ for pulleys with three different radii at rotation speeds from 2000 to 6000 RPM, at intervals of 1000 RPM.

experiments. Good agreement between the calculated and experimental temperature data proves the high trustworthiness of this thermal model. A maximum difference below 20% of the absolute elevated temperature is acceptable for industrial use and relevant aviation standards.

7.4.4 Temperature Plot Verification

Thus far the belt temperature is provided and validated. The next step is to prove the accuracy of the temperature distributions predicted for each pulley by experiments to achieve the temperature prediction of the entire belt drive system. The thermal model output is the heat flux flowing into each pulley and the belt, which determines the temperature distribution of every component. The belt temperature is uniform, and the accuracy of the prediction proven in Section 7.4.3. The temperature plot for a pulley under the targeted conditions is examined here. Figure 7.4 indicates that Zone 2 has the largest radial dimension compared with Zones 1 and 3, and it is reasonable to compare the temperature distribution of Zone 2 for each pulley to validate the accuracy of the thermal model. Figure 7.18 displays the temperatures from both the thermal model and experiment at five locations along the radial direction in Zone 2 of the 108 mm pulley, indicating good agreement between the data from the calculation and experiments. The 4 °C maximum temperature difference demonstrates that the results provided by the model can meet related industrial standards.

There are some notable differences between the calculated and experimental values. The thermal model is calculated under ideal conditions, while there are some negligible factors in the experiments. Measurement tolerances inside the devices exist and are impossible to eliminate, even after averaging the experimental data from three repeated tests. Another factor is the tradeoff between numerical accuracy and computational cost. The length of the elemental edge, as the smallest mesh in the turbulence simulation, could not become

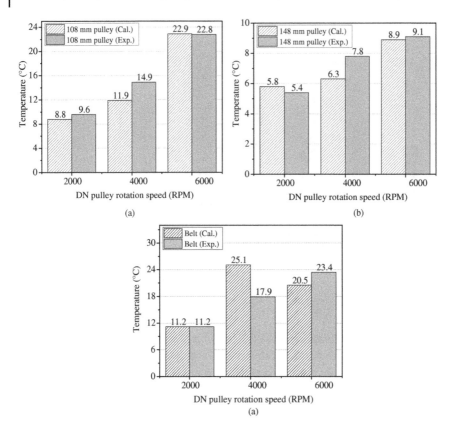

Figure 7.17 Calculated (Cal.) and experimental (Exp.) temperature comparisons for (a) 108 mm pulley, (b) 148 mm pulley, and (c) belt within the system.

infinitely small due to the high cost of computational resources this would have demanded. Thus, the predicted results are slightly different from the experimental values. Nevertheless, the above figures demonstrate that the results of this model are accurate.

Most importantly, all predicted calculations in this model uses considerably less computational resources compared with existing methods. For each load case, the total calculation time for a general calculation is less than three seconds at a desktop equipped with an AMD 3950X CPU, exclusive of the precalculation time for those simulations under baseline conditions. With such a speed, real-time temperature calculations can be achieved. Conversely, the currently available technique can require as long as 17 hours [28] to perform a similar temperature calculation for a typical belt drive. Therefore, the time comparison proves that this thermal model not only provides accurate but also instant temperature distribution results for belt drive design and optimization.

7.5 Conclusion

This work presents an approach for computing temperature predictions for belt drive systems used in aviation applications. A system of balanced thermal equations based on

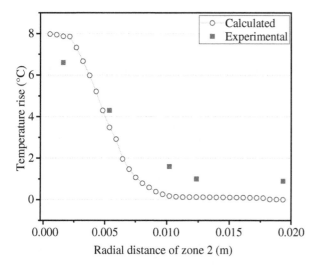

Figure 7.18 Calculated and experimental temperature rise comparison for Zone 2 for the 108 mm pulley at a rotation speed of 6000 RPM.

thermal behavior and exchange is used to import and modify precalculated numerical simulation results, providing precise real-time temperature information pertaining to irregular structures in a variety of load cases. To prove the high reliability of such a model, experiments under a wide range of operating conditions are conducted on a dynamometer. Comparisons between the calculated and experimental temperatures are made and discussed and good agreement noted, demonstrating the precision and consistency of this model. Therefore, it is possible to predict precise temperatures on the complex structures of transmission belt drive systems in helicopters under various operating conditions in only a few seconds, improving the life expectancy and safety margin for this type of aviation transportation.

References

1 Australian Transport Safety Bureau (2013). Reliability of the Robinson R22 helicopter belt drive system.

2 Gerbert, G. (1999). *Traction Belt Mechanics: Flat Belts, V-Belts, V-Rib Belts*. Chalmers University of Technology.

3 Manin, L., Liang, X., and Lorenzon, C. (2014). Power losses prediction in poly-v belt transmissions: application to front engine accessory drives. In: *International Gear Conference. 2014*, Lyon (26–28 August 2014), 1162–1171. Elsevier.

4 Balta, B., Sonmez, F.O., and Cengiz, A. (2015). Speed losses in V-ribbed belt drives. *Mech. Mach. Theory* 86: 1–14.

5 Silva, C.A., Manin, L., Andrianoely, M.-A. et al. (2019). Power losses distribution in serpentine belt drive: modelling and experiments. *Proc. Inst. Mech. Eng. Part D J. Automob. Eng.* https://doi.org/10.1177/0954407018824943.

6 Silva, C.A.F., Manin, L., Rinaldi, R.G. et al. (2017). Modeling of power losses in poly-V belt transmissions: hysteresis phenomena (standard analysis). *J. Adv. Mech. Des. Syst. Manuf.* https://doi.org/10.1299/jamdsm.2017jamdsm0085.

7 Silva, C.A.F., Manin, L., Rinaldi, R.G. et al. (2018). Modeling of power losses in poly-V belt transmissions: hysteresis phenomena (enhanced analysis). *Mech. Mach. Theory* 121: 373–397.

8 Bertini, L., Carmignani, L., and Frendo, F. (2014). Analytical model for the power losses in rubber V-belt continuously variable transmission (CVT). *Mech. Mach. Theory* 78: 289–306.

9 Qin, Z., Cui, D., Yan, S., and Chu, F. (2016). Hysteresis modeling of clamp band joint with macro-slip. *Mech. Syst. Sig. Process.* 66–67: 89–110.

10 Bents, D.J. (1981). Axial Force and Efficiency Tests of Fixed Center Variable Speed Belt Drive. *SAE Tech. Pap. Ser.* https://doi.org/10.4271/810103.

11 Julió, G. and Plante, J.S. (2011). An experimentally-validated model of rubber-belt CVT mechanics. *Mech. Mach. Theory* 46: 1037–1053.

12 Merghache, S.M. and Ghernaout, M.E.A. (2017). Experimental and numerical study of heat transfer through a synchronous belt transmission type AT10. *Appl. Therm. Eng.* 127: 705–717.

13 Fortunato, G., Ciaravola, V., Furno, A. et al. (2015). General theory of frictional heating with application to rubber friction. *J. Phys. Condens. Matter* https://doi.org/10.1088/0953-8984/27/17/175008.

14 Mackerle, J. (2004). Rubber and rubber-like materials, finite-element analyses and simulations, an addendum: a bibliography (1997–2003). *Modell. Simul. Mater. Sci. Eng.* 12: 1031–1053.

15 Phan, D. and Kondyles, D. (2003). Rotor Design and Analysis; A Technique Using Computational Fluid Dynamics (CFD) and Heat Transfer Analysis. *SAE Tech Pap. Ser.* https://doi.org/10.4271/2003-01-3303

16 Ramzan, M., Riasat, S., Kadry, S. et al. (2019). Numerical simulation of 3D condensation nanofluid film flow with carbon nanotubes on an inclined rotating disk. *Appl. Sci.* 10: 168.

17 McPhee, A.D. and Johnson, D.A. (2008). Experimental heat transfer and flow analysis of a vented brake rotor. *Int. J. Therm. Sci.* 47: 458–467.

18 Yan, H.B., Feng, S.S., Yang, X.H., and Lu, T.J. (2015). Role of cross-drilled holes in enhanced cooling of ventilated brake discs. *Appl. Therm. Eng.* 91: 318–333.

19 Belhocine, A. and Bouchetara, M. (2012). Thermal behavior of full and ventilated disc brakes of vehicles. *J. Mech. Sci. Technol.* 26: 3643–3652.

20 Talati, F. and Jalalifar, S. (2009). Analysis of heat conduction in a disk brake system. *Heat Mass Transfer* 45: 1047–1059.

21 Reddy, S.M., Mallikarjuna, J.M., and Ganesan, V. (2008). Flow and Heat Transfer Analysis of a Ventilated Disc Brake Rotor Using CFD. *SAE Tech Pap. Ser.* https://doi.org/10.4271/2008-01-0822

22 Wu, Y., Wu, M., Zhang, Y., and Wang, L. (2014). Experimental study of heat and mass transfer of a rolling wheel. *Heat Mass Transfer* 50: 151–159.

23 Moradi-Dastjerdi, R., Payganeh, G., and Tajdari, M. (2018). Thermoelastic analysis of functionally graded cylinders reinforced by wavy CNT using a mesh-free method. *Polym. Compos.* 39: 2190–2201.

24 Moradi-Dastjerdi, R. and Payganeh, G. (2017). Transient heat transfer analysis of functionally graded CNT reinforced cylinders with various boundary conditions. *Steel Compos. Struct.* 24: 359–367.

25 Safaei, B., Moradi-Dastjerdi, R., Qin, Z. et al. (2019). Determination of thermoelastic stress wave propagation in nanocomposite sandwich plates reinforced by clusters of carbon nanotubes. *J. Sandwich Struct. Mater.* https://doi.org/10.1177/1099636219848282.

26 Safaei, B., Moradi-Dastjerdi, R., Behdinan, K. et al. (2019). Thermoelastic behavior of sandwich plates with porous polymeric core and CNT clusters/polymer nanocomposite layers. *Compos. Struct.* https://doi.org/10.1016/j.compstruct.2019.111209.

27 Moradi-Dastjerdi, R. and Behdinan, K. (2019). Thermoelastic static and vibrational behaviors of nanocomposite thick cylinders reinforced with graphene. *Steel Compos. Struct.* 31: 529–539.

28 Wurm, J., Fitl, M., Gumpesberger, M. et al. (2016). Novel CFD approach for the thermal analysis of a continuous variable transmission (CVT). *Appl. Therm. Eng.* 103: 159–169.

29 Wurm, J., Fitl, M., Gumpesberger, M. et al. (2017). Advanced heat transfer analysis of continuously variable transmissions (CVT). *Appl. Therm. Eng.* 114: 545–553.

30 Mukunthan, S., Vleugels, J., Huysmans, T. et al. (2019). Thermal-performance evaluation of bicycle helmets for convective and evaporative heat loss at low and moderate cycling speeds. *Appl. Sci.* 9: 3672.

31 Shiming, Y. and Wenquan, T. (1998). *Heat Transfer*, 4e. Beijing: Higher Education Press.

32 Barton, D.E., Abramovitz, M., and Stegun, I.A. (1965). Handbook of mathematical functions with formulas, graphs and mathematical tables. *J. R. Stat. Soc. Ser. A* 128: 593.

33 Liu, X. and Behdinan, K. (2020). Analytical-numerical model for temperature prediction of a Serpentine Belt drive system. *Appl. Sci.* 10: 2709.

8

An Efficient Far-Field Noise Prediction Framework for the Next Generation of Aircraft Landing Gear Designs

Sultan Alqash and Kamran Behdinan

Advanced Research Laboratory for Multifunctional Lightweight Structures (ARL-MLS), Department of Mechanical & Industrial Engineering, University of Toronto, Toronto, Canada

8.1 Introduction and Background

In the early 1950s, aircraft noise was considered as one of the environmental noise impacts when turbo-jet engines were used to power aeroplanes. Thus, the aircraft noise issue was dominated by jet noise (i.e. a propulsive noise). However, in the early 1970s, the airframe of the aircraft itself has become the major source of noise after development of quieter engines (e.g. turbofan engines with high bypass ratios) [1]. Landing gears (LGs) are among the main noise sources generated during the approach-to-land phase of flight compared with the other airframe components [2, 3]. Since the number of people who live near airports is rapidly increasing, the Federal Aviation Administration (FAA) has imposed strict regulations on the aircraft noise level [4, 5]. Both US and European governments have set a target for 2020 to minimize the noise level by 50% near airport areas [6]. By 2050, the overall noise level of the aircraft should be further reduced by 65% [7]. A summary of aircraft noise history with the future targeted percentage of noise level reduction is depicted in Figure 8.1.

All the above-mentioned demands created pressure on academic communities and industries to develop better noise prediction and reduction techniques [8–14]. For a commercial aircraft, the noise generated from the airframe has received much attention in recent years. During the design stage of the LG, the aerodynamic aspect of the components is not addressed properly. As a result, many components are exposed to the air flow, and an aerodynamicnoise is generated. Basically, the main noise source mechanism is initiated due to the flow separation over the LG components which leads to multiple unsteady wakes that are characterized with intense flow instability perturbations [15].

In the literature, there are three distinct approaches to study the flow field and associated acoustic field around the LG within certain constraints or limitations. These are experimental, empirical/semi-empirical, and numerical approaches. Both full-scale and small-scale models for different types of LG such as two-wheel nose landing gear (2-NLG) and two-wheel, four-wheel, and six-wheel main landing gear (MLG) were tested in closed and open wind tunnels [9, 16–20]. The fluctuating forces produced due to the unsteady surface pressure around different LG components generate a wide noise frequency spectrum. Therefore, the LG has been observed to generate noise that has broadband spectral

Advanced Multifunctional Lightweight Aerostructures: Design, Development, and Implementation,
First Edition. Kamran Behdinan and Rasool Moradi-Dastjerdi.

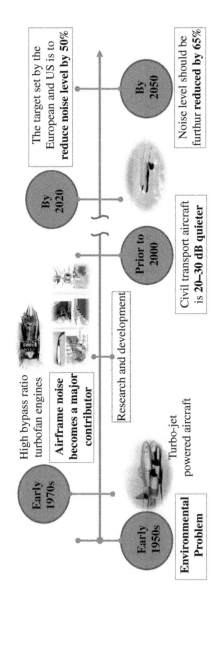

Figure 8.1 History of aircraft noise sources and future plan of noise level reduction.

characteristics [19]. In addition, the LG noise mechanism tends to be even more complex due to the close proximity between the components, which leads to a more complex flow field interaction [21].

Thus, for the next generation of LGs the low noise level should be one of the design objectives to assess the noise characteristics of different LG models. In general, the noise prediction through experiments is time-consuming and expensive [22, 23]. Developing reliable noise prediction tools is important to accelerate the design process. Therefore, new methods of noise reduction and control could be investigated to develop a low-noise LG. The key aspects of the prediction tool are to be reasonably accurate and easy to implement. There are some semi-empirical prediction methods [24–26] that have been utilized to estimate the overall noise of the LG. However, these semi-empirical models lack detailed description on the far-field noise spectrum. Therefore, they have limited applications as design tools. In contrast, the use of a physics-based approach, where the governing equations are numerically solved, will increase the amount of details in the noise prediction. In the early twenty-first century, the numerical approach has become an important tool to predict the aerodynamic noise.

Therefore, LG noise attenuation is critical to achieve better overall aircraft noise reduction. There are different techniques for noise reduction through both passive controls such as fairings/caps to streamline the bulky components as well as active controls such as actuators to control the vortex shedding [27–32]. To decide which noise reduction could be implemented on the LG, it first requires a quick estimate of its far-field noise to be obtained. Knowing the nature of flow-induced noise mechanisms and their locations, one can sufficiently target them and mitigate their associated noises [33–42]. For that reason, the focus of this research is to develop a tool that accelerates the far-field noise prediction of LG. This tool is suitable during the design process instead of building a large experimental setup or running costly three-dimensional (3D) numerical simulations.

8.1.1 Numerical Landing Gear Aeroacoustics

Most of previous numerical studies for LGs utilized the Ffowcs Williams and Hawkings (FW-H) equation for the prediction of its noise in far-field. The FW-H formulation considers both impermeable walls and permeable interior surfaces as integration surfaces. Unlike the Kirchhoff method, the latter surface can be placed in a nonlinear region [43]. The FW-H method has been shown to be numerically superior to Kirchhoff integral formulation [44] for many applications. Van Mierlo et al. [45] studied the influence of the MLG inclination angle on the noise level using ANSYS Fluent. However, they did not include the contribution of the quadrupole sources corresponding to the volume integral due to its high computational cost. Thus, the selection of the surface always coincides with the wall of the buff body. The same approach was used by Long et al. [39] to study the far-field noise of a simplified NLG. They noticed that the directivity of the noise is a dipole like source. Another method based on computational fluid dynamics (CFD) and the FW-H equation with a penetrable surface has advanced considerably. For instance, Lockard et al. [46] compared the results and noticed that the noise predicted based on a solid surface agreed well with the results of near-field CFD, and that noise predicted based on a penetrable surface could be corrupted by the pseudo-sound created on closing the FW-H surface in the wake region.

Also, Spalart et al. [47] and Sanders et al. [48] argued that by calculating the noise based only on a solid surface would lead to inaccurate results even at Mach numbers as low as 0.115. The results revealed a typical discrepancy of 3 dB compared with the permeable surface. Similarly, Souliez et al. [49] predicted the noises of two geometric configurations of a four-wheel LG without and with lateral struts using the FW-H equation. The flow data of near-field were collected by an unstructured finite volume low-order solver. The FW-H equation was solved by calculating surface integrals over a permeable surface far from the LG and over the LG surface. Porous FW-H surface results matched well with experimental data compared with results based on the solid FW-H surface due to not including all the pressure fluctuations. Therefore, it was found that quadrupole effects could play a key role in overall near-field sound level even at moderate Mach numbers. There are some problems observed in the available numerical aeroacoustic results, where the noise spectra of the 3D LG model have discrepancies compared with the measurements [49–52]. For example, Liu et al. [53] applied a CFD/FW-H hybrid technique to study the aerodynamic noise of a generic 2-NLG. Unsteady flow field was calculated using a compressible Navier–Stokes (N-S) solver based on a fully structured grid and high-order finite difference approaches. The results correspond to the 2-NLG LAnding Gear nOise database for CAA validatiON (LAGOON) model. The power spectral density (PSD) spectrum results with respect to frequency showed a large discrepancy between measurement and computations especially in the mid-to-high frequency range from 2 kHz and beyond. It also indicted a strong dependency on the grid refinement such that the increase of cell numbers from 3.5 to 15.7 million matched the measurement data.

Moreover, in the same study, the far-field results of the flyover and sideline measurements were compared with the numerical results of different solvers such as in-house code (Soton-CAA) and ANSYS Fluent. The results of the 3D LG model showed a lack of accuracy where the PSD spectrum drops off dramatically at approximately 1 kHz and beyond. Refined grid and baseline calculations were performed on two supercomputing groups of IRIDIS3 and Spitfire, available at the University of Southampton. Also, a parallel computing approach based on decomposing the domain and using the Message Passing Interface (MPI) libraries was utilized. It can be noted that the 3D simulations were very time consuming and still the results were not accurate enough.

8.1.2 Problem Statement

Based on the state-of-the-art, one can see that there is a need to develop a new LG noise prediction tool that captures enough details instead of relying on empirical constants. Thus, it can be utilized as a quick design tool. The available empirical or semi-empirical models are dependent on the calibration of the experimental results, which makes them insufficient in predicting the far-field noise of a new LG design. Overall, the results based on these empirical models always have a lack of details in the noise spectra.

Noise generated aerodynamically occurs first as the result of moving bodies. Due to this movement and resulting flow unsteadiness, the pressure fluctuations are generated, and then the surrounding medium (e.g. air molecules at rest) is excited and radiates noise. It is called impulsive noise. The second noise is generated due to flow turbulence. The aeroacoustic noise is generally composed of broadband noise resulting from the turbulence and tonal peaks generated by impulsive noise [54]. Therefore, the quadrupole source has to be included by using the FW-H porous surface method in order to improve the accuracy of the predicted noise.

8.2 Modeling and Numerical Method

8.2.1 Hybrid Computational Aeroacoustic Method

In this chapter, all the simulations are performed using the hybrid computational aeroacoustic (CAA) method. Basically, the computational procedure is divided into two steps. Near-field flow computation is decoupled from the computation of acoustics at far-field. In the first step, the fluid dynamics governing laws are numerically solved around various bluff bodies on a computational grid. The integral solutions of N-S equations are numerically discretized and solved at each domain cell. All details about time-dependent flow information near the noise source are collected. This process of near-field data sampling is performed by specifying the FW-H integration surfaces that surround the acoustic sources. As mentioned earlier, there are two methods such as solid/impermeable surface (on-body) and porous/permeable surface (off-body). In the second step, the obtained CFD data from the first step are used as inputs to solve the FW-H equation and predict acoustic pressure at far-field observer(s). Figure 8.2 describes the hybrid CAA method which is a two-step approach and near-field to far-field distance criterion.

The near-field acoustic pressure, p′, decays with the inverse square of the receiver distance, R, (i.e. p′ ∝ 1/R²). However, in the far-field region the p′ decays as the inverse of the R (i.e. p′ ∝ 1/R). Therefore, the criterion separating the near-field and far-field regions is wavelength, λ. In other words, when the R becomes sufficiently larger than the λ, only the far-field remains. In practical applications, the far-field condition is met about one λ away from the source (e.g. between a few centimeters to a few meters in the audible range 20 Hz–20 kHz). One of the advantages of this method is that small perturbations in acoustic do not have to be solved up to far-field observers. Based on this, the computational cost is dramatically reduced in comparison to the direct numerical simulation (DNS). In the following, more details are provided for the numerical methods that are utilized to solve for the near-field flow and far-field acoustic domains.

8.2.2 Near-Field Flow Numerical Method

The fluid flow motion of any Newtonian fluid such as air is governed by the N-S equations. The N-S equations are a combination of two main equations in fluid mechanics, namely: mass conservation (i.e. continuity equation) and momentum conservation (e.g. forces balance, $\sum \mathbf{F}$, using Newton's second law, $\sum \mathbf{F} = m\mathbf{a}$, in which m [kg] is mass and $\mathbf{a}$ [m/s²] is the acceleration) that describe the fluid as a continuum material and the system is conserved. In this study, the two-dimensional (2D) flow domain is considered, where its governing equation in the case of incompressible condition (constant density, $\rho = $ constant) is written as [41]:

Continuity equation:

$$\frac{\partial u}{\partial x} + \frac{\partial v}{\partial y} = 0 \tag{8.1}$$

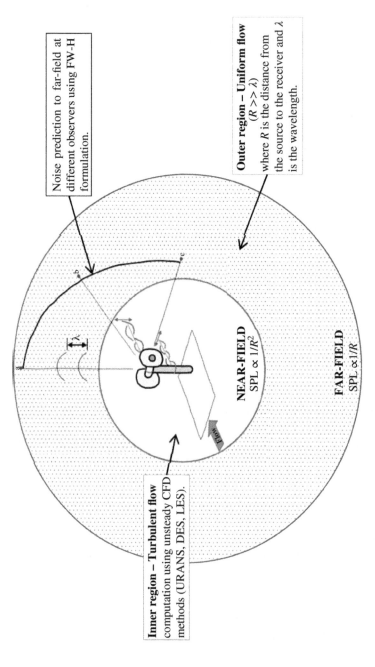

Noise prediction to far-field at different observers using FW-H formulation.

Outer region – Uniform flow $(R >> \lambda)$ where R is the distance from the source to the receiver and λ is the wavelength.

NEAR-FIELD SPL $\propto 1/R^2$

FAR-FIELD SPL $\propto 1/R$

Flow

Inner region – Turbulent flow computation using unsteady CFD methods (URANS, DES, LES).

Figure 8.2 Illustration of the hybrid CAA computational approach and near-field to far-field distance criterion.

Momentum equations:

x-component:

$$\rho\left(\frac{\partial u}{\partial t} + u\frac{\partial u}{\partial x} + v\frac{\partial u}{\partial y}\right) = -\frac{\partial P}{\partial x} + \mu\left(\frac{\partial^2 u}{\partial x^2} + \frac{\partial^2 u}{\partial y^2}\right). \tag{8.2}$$

y-component:

$$\rho\left(\frac{\partial v}{\partial t} + u\frac{\partial v}{\partial x} + v\frac{\partial v}{\partial y}\right) = -\frac{\partial P}{\partial y} + \mu\left(\frac{\partial^2 v}{\partial x^2} + \frac{\partial^2 v}{\partial y^2}\right) \tag{8.3}$$

The above equations are usually written in a more concise form using the vector notation as:

$$\nabla.\mathbf{u} = 0 \tag{8.4}$$

$$\rho\left(\frac{\partial \mathbf{u}}{\partial t} + (\mathbf{u}.\nabla)\,\mathbf{u}\right) = -\nabla P + \mu\nabla^2\mathbf{u} \tag{8.5}$$

where P is pressure, $\nabla = \frac{\partial}{\partial x}\widehat{i} + \frac{\partial}{\partial y}\widehat{j}$ is the gradient, ∇^2 is the Laplace operator, ρ is density, and $\mathbf{u}$ is the velocity vector (e.g. u and v are the components of velocity along the x and y directions, respectively). Thus, one can solve the above equations providing the boundary conditions based on the problem (discussed in detail in the following sections). However, solving Eqs. (8.4) and (8.5) by the DNS method leads to prohibitive calculation time; instead, the large eddy simulation (LES) model is applied in this chapter.

8.2.3 Far-Field Acoustic Numerical Method

8.2.3.1 Acoustic Analogies Formulation

First, Lighthill formulated the acoustic analogy wave equation where the source derived by comparing exact fluid motion equations with sound propagation equations for a medium at rest is [41]:

$$\frac{1}{c_0^2}\frac{\partial^2 \rho'}{\partial t^2} - \nabla^2\rho' = \frac{\partial^2 T_{ij}}{\partial x_i\partial x_j} \tag{8.6}$$

$$T_{ij} = \rho v_i v_j + P_{ij} - c_0^2\rho'\delta_{ij} \tag{8.7}$$

where $(\rho' = \rho - \rho_o)$ indicates density perturbation, T_{ij} is Lighthill's stress tensor, $P_{ij} = p'\delta_{ij} - \tau_{ij}$ refers to the compressive stress tensor in which p' is fluctuating pressure and τ_{ij} is the total stress tensor equal to $\sigma_{ij} - p\delta_{ij}$, where σ_{ij} refers to the residual stress tensor. Note that v_i and v_j indicate the components of velocity along the x_i and x_j directions, respectively. Here, c_o is the velocity of sound in fluid at rest (i.e. speed of sound). Thus, Eq. (8.7) can also be written as $T_{ij} = \rho v_i v_j - \sigma_{ij} + (p' - c_0^2\rho')\,\delta_{ij}$.

Secondly, Curle developed Curle's analogy by expanding the theory of Lighthill taking into account a rigid surface [33]; after that, the theory was further generalized by taking into account a rigid object in arbitrary motion as [42]:

$$\left(\frac{\partial^2}{\partial t^2} - c_0^2\frac{\partial^2}{\partial x_i^2}\right)(H\,(g_s)\,\rho') = \frac{\partial^2}{\partial x_i\partial x_j}\,(T_{ij}H\,(g_s)) - \frac{\partial}{\partial x_i}\,(F_i\delta\,(g_s)) + \frac{\partial}{\partial t}\,(Q\delta\,(g_s))$$

$$\tag{8.8}$$

$$F_i = \left(P_{ij} + \rho v_i\left(v_j - u_j\right)\right)\frac{\partial g_s}{\partial x_j} \tag{8.9}$$

$$Q = \left(\rho_0 u_i + \rho\left(v_i - u_i\right)\right) + \frac{\partial g_s}{\partial x_i} \tag{8.10}$$

in which g_s represents body surface as a function of $g_s(x, t) = 0$; where $g_s > 0$ and $g_s < 0$ refer to outside and inside of the rigid body, respectively. The function of g_s describes the integration surface moving at speed u_j and is a function of time; therefore, it always encompasses a moving source region. The right-hand side of Eq. (8.8) contains source terms; the physical meaning of the first term shown explicitly in Eq. (8.7) is unsteadiness inside the fluid stated in the quadrupole source term as a Reynolds stress and is known as the turbulence-induced term, the second term is dipole source due to boundary dilatation, and the third term is monopole source modeled from surface fluctuating stresses. The last two terms shown in detail in Eqs. (8.9) and (8.10) describe loading and thickness noises, respectively. The differential form of FW-H shown in Eq. (8.8) could be expressed in terms of fluctuating pressure based on the following relationship $\left(p' = c_o^2\rho'\right)$ which shows the isentropic relation. There are three important functions used in the above equations namely Kronecker delta δ_{ij}, Heaviside $H(g_s)$, and the derivative of Heaviside function $H'(g_s) = \delta(g_s)$ is Dirac delta where each of them can be defined as follows:

$$\delta_{ij} = \begin{cases} 1, & i = j \\ 0, & i \neq j \end{cases}, H\left(g_s\right) = \begin{cases} 1, & g_s > 0 \\ 0, & g_s < 0 \end{cases}, \text{ and } \delta\left(g_s\right) = \begin{cases} 1, & g_s = 0 \\ 0, & g_s \neq 0 \end{cases}$$

Farassat's formulation 1A gives the equation of FW-H analogy as the following integral forms [13]:

$$4\pi P_{\text{Thickness}} = \int_{g_s=0}\left[\frac{\rho_0\dot{v}_n}{r\left(1 - M_r\right)^2} + \frac{\rho_0 v_n\hat{r}_i\dot{M}_i}{r\left(1 - M_r\right)^3}\right]_{\text{ret}} dS + \int_{g_s=0}\left[\frac{\rho_0 c_0 v_n\left(M_r - M^2\right)}{r\left(1 - M_r\right)^3}\right]_{\text{ret}} dS \tag{8.11}$$

$$4\pi P_{\text{Loading}} = \int_{g_s=0}\left[\frac{\dot{p}\cos\theta}{c_o r\left(1 - M_r\right)^2} + \frac{\hat{r}_i\dot{M}_i p\cos\theta}{c_o r\left(1 - M_r\right)^3}\right]_{\text{ret}} dS +$$
$$\int_{g_s=0}\left[\frac{p\cos\theta}{r^2\left(1 - M_r\right)} + \frac{\left(M_r - M^2\right)p\cos\theta}{r^2\left(1 - M_r\right)^3}\right]_{\text{ret}} dS \tag{8.12}$$

$$4\pi P_{\text{Quadrupole}} = \frac{\partial^2}{\partial x_i \partial x_j}\int_V\left[\frac{T_{ij}}{r\left|1 - M_r\right|}\right]_{\text{ret}} dV \tag{8.13}$$

where $1 - M_r$ is the Doppler factor, $r = |x - y|$ indicates the distance between the observation point of acoustic, x, and the point field where sound is generated in the flow, y, and all terms are calculated in retarded time $t_{\text{ret}} = t - r/c_0$. A volume integration is required for the turbulence term associated with the quadrupole source [55]. Thus, a permeable/porous FW-H equation was formulated by establishing a virtual permeable surface. It modifies the assumption of the function surface (g_s) that normal fluid velocity of (v_n) and normal rigid body velocity (u_n) are equal $(v_n = u_n)$ to be unequal $(v_n \neq u_n)$. By doing so, the calculations of loading noise and thickness on permeable surfaces (e.g. control surface or inner-cell) include turbulence-induced noise inside the surface [56, 57]. The first attempt to account for the quadrupole noise sources appearing in FW-H formulation was initiated while developing the Kirchhoff–FWH method [58]. The main idea behind that step is to take advantage of both approaches. This method was tested for predicting hovering rotor noise. In the original FW-H equation, computation of the quadrupole term needs volume integration in regions near the control surface and body which leads to a high computational cost. For that reason, in most of the studies [39, 45] the quadrupole source is ignored. To avoid these problems, a new formulation (permeable/porous FW-H) was developed and examined for a flow simple case past a circular cylinder [59].

8.2.3.2 Ffowcs Williams and Hawkings Equation

The FW-H equation [42] is an inhomogeneous wave equation obtained from the exact rearrangement of N-S equations. The FW-H equation was previously shown in terms of density in Eq. (8.8); however, it could be stated in terms of far-field acoustic pressure ($p' = p - p$), as follows:

$$
\frac{1}{c_0^2}\frac{\partial^2 p'}{\partial t^2} - \nabla^2 p' = \frac{\partial^2}{\partial x_i \partial x_j} T_{ij} H\left(g_s\right) - \frac{\partial}{\partial x_i}\left[P_{ij} n_j + \rho v_i\left(v_j - u_j\right)\right]\delta\left(g_s\right)
$$
$$
+ \frac{\partial}{\partial t}\left[\rho_0 u_i + \rho\left(v_i - u_i\right)\right]\delta\left(g_s\right)
$$

(8.14)

where n_j is the outwards unit normal vector to surface $g_s = 0$.

The solution of differential FW-H, Eq. (8.14), is achieved by the free-space Green function $(G(x, y, t, \tau) = \frac{\delta(g)}{4\pi r}$, where $g = \tau - t + \frac{|x-y|}{c_0}$, retarded time $\tau = t - r/c_0$ and distance to the observer $r = |x - y|$). The complete solution consists of volume and surface integrals. Surface integrals reveal the impacts of dipole and monopole acoustic sources and partially quadrupole sources. In this chapter, dipole sources are generated due to the force fluctuations around the bluff bodies, whereas the monopole sources are generated due to flow displacement. The volume integral reveals quadrupole sources in the outside region of the source surface. In Fluent, the volume integral is dropped to reduce the computational efforts. However, when a permeable/porous surface is located at a certain distance from the bluff body, the integral solution of Eq. (8.8) includes the contribution of quadrupole source within the region enclosed by that permeable surface. Therefore, different locations for the porous surface are examined. When the surface is placed on the solid body itself, only the unsteady surface pressure is recorded and employed in the FW-H solver to find the far-field acoustic pressure. However, when the surface is located away from the bluff body, all the pressure, density, and velocity are saved and then used by the FW-H solver to account for the quadrupole source. In the calculation of sound pressure by the FW-H solver, the forward-time projection method is applied to consider the time delay between reception and emission times. Once the unsteady solution of 2D near-field flow becomes fully developed (e.g. the flow reaches a statistically steady state), the storage of the time history of flow quantities on the predefined surface(s) is initiated. Then, the simulation continues for more time steps until enough sample is recorded. To compensate for the missing flow data in the third dimension, the 2D CFD is repeated along a certain length. This length is known as a source correlation length (SCL). In this research, the SCL is assumed to be very short, so that the primary vortex shedding can be extracted. The assumption of this SCL parameter is based on maintaining the flow coherence along the span direction. To ensure that the flow is fully correlated along the span, the SCL is made to be smaller than bluff body characteristic length. Figure 8.3a shows the different locations of impermeable surface (Solid) versus permeable surface(s) (Circ-domain and Rect-domain) and Figure 8.3b is a schematic of how the SCL is assumed to map the flow data in the third direction. By assuming that the SCL is almost equal or smaller than the characteristic length, D, of the bluff body, the acoustic source compactness condition (wavelength, $\lambda \gg D$) is satisfied. Thus, the difference between emission times of points along the span are considered negligible compared with pressure fluctuation period.

After specifying the location of the source and then assuming the SCL, further analysis is carried out to calculate the far-field acoustic pressure at different receiver locations. Finally, the timeline of acoustic pressure is post-processed to obtain acoustic directivity and spectral characteristics of the sources. The fast Fourier transform (FFT) tool in Fluent is utilized for transferring sound pressure signal from the time domain to the frequency domain. Then, both the PSD and sound pressure level (SPL) are plotted. After that, these data are saved in separate files for further analysis. These files are utilized in the calculation of the total noise level based on the summation of the signal noise of each 2D case using a Microsoft Excel spreadsheet.

8.2.4 The Motivation of the Multiple Two-Dimensional Simulations Method

The idea of using the CFD data of multiple 2D simulations instead of performing a costly 3D computation originated after exploring that there are different fluid measurement techniques. The most popular measurements are hot-wire anemometry and Particle Image Velocimetry (PIV). They are employed to find velocity fluctuating regions. These measurements can serve to obtain some information about the noise sources and their mechanisms. In addition, the unsteady pressure transducers that are located on the model surface have also facilitated the detection of the sound production region. The radiated noise is then measured using the microphones that are positioned around the model. After that, the directivity pattern can be determined by locating these microphones at different angles. The idea was shaped out after reading the experimental studies conducted by Manoha et al. [60] on the LAGOON NLG. PIV measurements were used to collect flow quantities inside the French national aerospace research center (ONERA F2) aerodynamic closed wind tunnel. Interestingly, it can be noted that the PIV planes were positioned at different levels around the wheels and the main struts. This motivated the implementation of the idea of using multiple 2D CFD data for the calculation of LG noise at far-field. LAGOON NLG is chosen as the first candidate problem to validate the methodology. In the following sections, more details of the proposed approach framework, implementations, and advantages are highlighted.

8.2.5 New Approach for Noise Calculation at the Far-Field of LG

Here, a 2D simulation method based on physics is developed and validated by existing 3D numerical and experimental data. The developed method is more efficient and effective in compromising the accuracy and computational cost (i.e. the required time and CPU memory) of the prediction of far-field noise in complex structures such as LG. Noise prediction at

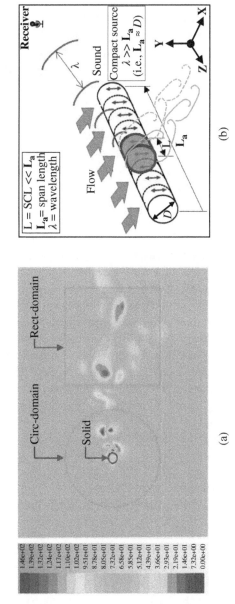

Figure 8.3 Flow past a circular cylinder at Re $= 9.0 \times 10^4$: (a) different acoustic source locations on a velocity magnitude contour image; and (b) schematic of the SCL assumption.

far-field relies on the flow data of multiple 2D cross-sections at the near-field of LG. These cross-sections are adopted from various locations depending on LG complexity, such as the number of components, and their sizes and orientations. FW-H acoustic analogy is applied to calculate noise at far-field. For the calculation of noise at far-field using the obtained flow data of 2D cross-sections at near-field, a SCL is assigned to compensate for the missing CFD data along the span. It was found that flow was completely correlated only along a few body diameters in different shapes of bluff body at moderate to high Reynolds number (Re) [61–64]. In the developed method, SCL is in compliance with the computed local Reynolds number, Re^*, using each cross-sectional length, L_a. Furthermore, all major changes in the geometry of the cross-section which alter flow properties are considered in SCL. In aerodynamic flows, in addition to the 3D nature of vortex shedding, two irregularity types affect coherence level [65]. For example, cylinder or flow irregularity disturbs the homogeneity of flow. Cylinder irregularity can be further divided into section, surface and spanwise irregularities. Flow irregularity can be categorized into variable local radius, axis deviation, and end effects (more details are given in [65]). The proposed method has the following steps for both flow and acoustic domains [66].

Flow domain setup

1) Dividing the developed 3D LG model into several slices with different dimensions and configurations. For bluff bodies with long aspect ratios (i.e. the ratio of length to diameter is higher than three), mainly located at the upper parts of LG, one cross-section is enough to capture the behaviors of flow. Meanwhile, the lower LG part has bulky components with small aspect ratios; thus, the minimum of three horizontal cross-sections should be considered. For instance, one cross-section could be considered at the center plane of the wheels-axle area and the other two could symmetrically be considered at the lower and upper portions of wheels. These two cross-sections should be considered far from the lower and upper circumference zones of the wheels. They should also be considered far from the inner rim of the cavity to prevent flow separation points.
2) Separate modeling and sizing of the 2D calculation domain for each cross-section according to the corresponding L_a.
3) Discretizing each domain by a hybrid mesh, where a structured mesh is employed around the body in order to capture the boundary layer, and an unstructured mesh is applied to the whole domain.
4) Simulating each 2D domain separately by the CFD solver.
5) Extracting and saving all essential flow data at the near-field including unsteady surface pressure after establishing a statistically steady state as an acoustic calculation input in the following step.

Acoustic calculations

1) Specifying outer surfaces of each obtained cross-section as acoustic sources.
2) Locating a receiver for each single domain above the body at a distance based on the location of microphone in the performed test.
3) Performing simulations in multiple time steps to capture sufficient samples for a documented acoustic pressure signal.

4) For the calculation of noise at far-field based on 2D flow information, SCL is estimated according to the geometry and flow characteristics of the object as well as available empirical equations and experimental data.

5) Overall LG far-field noise is calculated by the summation of all noise contributions of each cross-section.

8.3   Implementation of the Multiple Two-Dimensional Simulations Method

The developed method is applied to LAGOON NLG, which is assumed as a benchmark problem in the second Benchmark Problems for Air frame Noise Computations workshop (BANC-II) [67]. The experimental data of LAGOON NLG [68] and 3D numerical results [69] are applied in the validation of the developed method. Computational domain size of each cross-section is independently determined according to the corresponding L_a (Figure 8.3). 2D CFD analysis is then performed on each cross-section to provide flow data at near-field. Then, total PSD is obtained as the summation of all noises propagated from each 2D cross-section. Among the advantages of the developed method is that CFD analyses of 2D cross-sections are individually performed. Therefore, the size of domain, process of mesh generation, and time of computation are significantly decreased in comparison with the corresponding 3D LG model.

Figure 8.4a illustrates the Z-axis locations of five different cross-sections and Figure 8.4b shows the shapes of those cross-sections in the 3D LAGOON NLG model. The cross-sections utilized in CFD simulations are described in the following [66]:

• The first plane is positioned on the main strut, so that the structures of flow and noises generated near vertical circular cylinders can be found.

• The second plane is assumed to be at the following subset cylindrical strut element with a smaller diameter. Therefore, this cross-section captures flow behavior variations.

• The third one is located at the upper rim of the outer cavity to capture the parameters of essential flow in the upper parts of both wheels.

• The fourth plane presents a more complicated cross-section at the center of the wheel-axle.

• Ultimately, the fifth plane is placed at the lower rim of the outer cavity. The shape of this cross-section and the third one is similar. However, the fifth cross-section lacks the circular part of the strut between the wheels.

These five main cross-sections are specified to obtain various flow behaviors and evaluate their associated noises (identified as Sec.1–Sec.5).

8.3.1   2D CFD Setup

Once the cross-section number is set for various configurations and components, dimensions are accordingly achieved by CAD software. To facilitate obtaining cross-section dimensions, the complete 3D LAGOON NLG is modeled using SolidWorks. Then, ANSYS Workbench is used to separately design the computational domain of flow for each

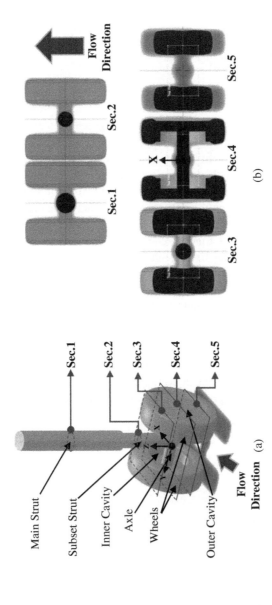

Figure 8.4 The 3D LAGOON NLG model with (a) Z-axis locations of the five planes and (b) 2D shapes of cross-sections resulting from the five planes.
Source: [66]. Licensed under CC BY 4.0.

cross-section. Boundary conditions and dimensions of 2D computational flow domain are shown in Figure 8.5. It is noteworthy that, in Sec.3 to Sec.5, half of the cros-section lengths, L_a, are employed to decrease domain size.

Figure 8.6 shows 2D computational domain mesh configurations, where, as mentioned earlier, the hybrid mesh method is applied. Unstructured mesh is employed to discretize the whole domain, while structured mesh is applied around each cross-section contour to obtain flow separation and boundary layer.

The flow conditions obtained from the anechoic open wind tunnel are applied in all simulations [60]. The following values are used in experiments: static pressure $P_0 = 9.67723 \times 10^4$ Pa, flow density $\rho_0 = 1.18$ kg/m^3, and inlet velocity $U_{in} = 78.99$ m/s, which shows $M = 0.23$. In all simulations, the LES turbulence model is applied. The implicit second-order scheme is utilized to obtain the solution in the time domain. The non-iterative time advancement scheme is applied to increase the accuracy and acceleration of the solution in simulations. In all simulations, Δt is considered as 5.0×10^{-6} seconds to guarantee numerical stability and convergence. In the LES model, the size of Δt is controlled by the time scale of the smallest resolved eddies. Therefore, the Courant–Friedrichs–Lewy number remains close to 1. At this Δt value, the maximum frequency is 100 kHz. Ensuring quasi-stationary flow behavior before obtaining source data for acoustic analyses is very important. Table 8.1 summarizes 2D cross-section locations, evaluated flow factors and the properties of their mesh. The Re* of each cross-section is obtained according to the related L_a. LAGOON NLG has an inner cavity with wheel diameter $D_w = 300$ mm and side-to-side spacing $S = 198$ mm [60]. In full 3D LG model, Re $= 1.56 \times 10^6$ is determined according to wheel diameter because it has the greatest dimension of all components. Consequently, flow behavior is in the upper transition regime. Here, the calculated Re* has various flow regimes from critical lower transition to upper transition, including supercritical.

It is noteworthy that in the definition of the aspect ratio, l_c and D_c denote the length and diameter of the cylinder, respectively, and are only obtained for Sec.1 and Sec.2. Both struts have greater aspect ratios compared with the wheel component. In other cross-sections also, each wheel has a cylindrical shape, however with a smaller aspect ratio than the upper struts and with different orientations with respect to incoming flow. Furthermore, because of inner and outer cavities, these wheels cannot be considered as smooth and continuous bluff bodies. Therefore, for Sec.3 and Sec.4 the aspect ratio is not calculated, as shown in Table 8.1.

8.3.2 Computation of Acoustic at Far-Field

The built-in 3D FW-H solver of ANSYS Fluent is applied to calculate acoustic noise at far-field. In each cross-section, outer surfaces are considered as acoustic sources. The sources obtained from each 2D flow domain at near-field are extended into the third direction. For FW-H surface integration, this extension is elongated along the spanwise direction to the assumed SCL. This implies that third direction data are identical to source information obtained from 2D CFD simulations. Ultimately, acoustic pressure at far-field is determined at receivers positioned above bluff bodies at a distance of 6 m. As the source zone is considered at the body solid surface, only the FW-H equation surface integral is calculated. Therefore, only the dipole acoustic source obtained from the fluctuations

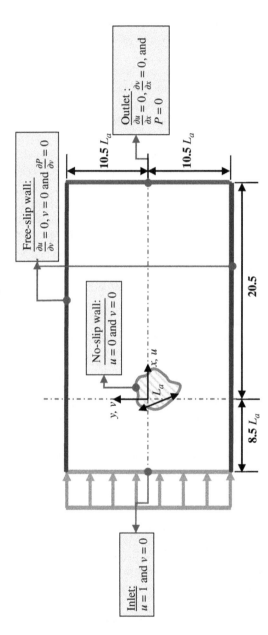

Figure 8.5 Schematic diagram of the size and boundary conditions of the 2D computational domain.

Table 8.1 Summary of mesh properties and flow factors for the considered 2D cross-sections in LAGOON NLG.

Case name	Z-axis locations L_z (m)	L_a (m)	Aspect ratio, l_c/L_a	Re*	Compactness condition parameters			SCL (m) ($L_a \ll \lambda$)	No. of elements
					St	f (Hz)	λ (m)		
Sec.1	+0.335	$D_{c1} = 0.071$	6.366	3.70×10^5		222.5	1.5	$5.0D_{c1} = 0.355$ $2.5D_{c1} = 0.178$ $0.8D_{c1} = 0.056$ [a]	8.11×10^4
Sec.2	+0.194	$D_{c2} = 0.055$	4.327	2.86×10^5		287.2	1.2	$5.0D_{c2} = 0.275$ $2.5D_{c2} = 0.138$ $1.58D_{c2} = 0.087$ [a]	5.00×10^4
					0.2 ¤				
Sec.3	+0.104	$D_{w3} = 0.214$	—	1.11×10^6		73.8	4.6	$2.5D_{w3} = 0.535$ $0.486D_{w3} = 0.104$ $0.107D_{w3} = 0.023$	9.71×10^4
Sec.4	0	$D_{w4} = 0.300$	0.293	1.56×10^6		52.7	6.5	$2.5D_{w4} = 0.751$ $0.347D_{w4} = 0.104$ $0.077D_{w4} = 0.023$	1.35×10^5
Sec.5	−0.104	$D_{w5} = 0.214$	—	1.11×10^6		73.8	4.6	$2.5D_{w5} = 0.535$ $0.486D_{w5} = 0.104$ $0.107D_{w5} = 0.023$	7.29×10^4

a) The values are calculated by empirical equation. D_{c1} and D_{w3} – the subscripts w and c denote non-circular cross-section and circular cross-section, respectively, while the numbers show case number. In this study, St is set to 0.2 based on Roshko [70] for all cross-sections.

Source: [66]. Licensed under CC BY 4.0.

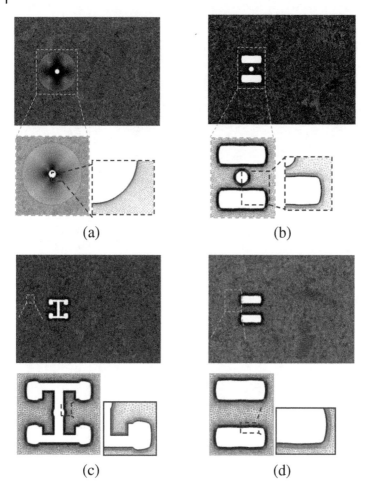

(a) (b)

(c) (d)

Figure 8.6 LG cross-section mesh configurations at (a) Sec.1 or Sec.2, (b) Sec.3, (c) Sec.4, and (d) Sec.5. Source: [66]. Licensed under CC BY 4.0.

of unsteady surface pressure is considered. It is noteworthy that near-field nonlinearity remains but the influence of the decrease of flow domain dimensionality on the calculation of noise at far-field is evaluated. The acoustic sources of the obtained cross-sections are assumed to be acoustically compact. This assumption is valid when $L_a \ll \lambda$, in which λ is the acoustic wavelength. To obtain an initial estimation of the Strouhal number (St), empirical equations were developed by Norberg [71] for a wide range of Re (i.e. 47–3.0×10^5) of flow past a stationary cylinder. For instance, in this work, Eq. (8.15) is applied for the estimation of St for Sec.1 and Sec.2 [71]:

$$St = 0.198 \left(1 - \frac{19.7}{Re^*}\right) \tag{8.15}$$

Curle's requirement [25] governs the magnitude of St when $St \ll 1/(2\pi M)$. For satisfying the condition of compactness for these two cross-sections with $M = 0.23$, it was considered that $St \ll 0.69$. One could expect that the results obtained at the main shedding tone

Figure 8.7 Z-axis locations of the lower part cross-section distances. Source: [66]. Licensed under CC BY 4.0.

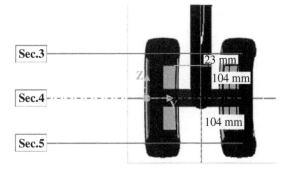

($St \approx 0.2$, from Eq. (8.15)) along with its harmonics can satisfy the requirement of Curle. Furthermore, the empirical equation of SCL estimation presented in [71] is also employed. As shown in Table 8.1, according to the evaluated Re^* for Sec.1 and Sec.2, Eq. (8.16) is applied for the calculation of SCL [71].

$$SCL = \left[1.4 \left(\frac{Re^*}{3.0 \times 10^5} \right)^{-2.7} \right] L_a \tag{8.16}$$

For the other cross-sections shown in Figure 8.7, SCLs are estimated from the geometric aspects of two wheels and the Re^* of each cross-section. The configurations of the lower part cross-sections of the considered 2-NLG are more complex than those of the upper parts. Thus, the considered SCL values avoid abrupt geometric variations such as curvatures and cavities. In Sec.3, for example, the value of $SCL1 = 2.5D_{w3}$ is equivalent to that of SCL2 for Sec.2. As can be seen in Table 8.1, the second $SCL2 = 0.486D_{w3}$ is determined according to the distance of Sec.3 and Sec.4 (104 mm). SCL3 is assumed to be $0.107D_{w3}$ to prevent surpassing the edges of inner cavities of wheels, which is considered to be 23 mm from Sec.3 (Figure 8.7).

For the calculation of total far-field noise, a similar number of acoustic pressure data is achieved for each cross-section. This corresponds to unsteady pressure at solid surface nodes of cross-sections. The total PSD of all noises propagated because of passing flow on those cross-sections is obtained from:

$$PSD_{Total} = 10 \log_{10} \left(\sum_{i=1}^{N} 10^{PSD_i/10} \right) \tag{8.17}$$

where N is the total cross-section number and i is the cross-section number index.

The total computational time of flow domain at near-field is $t = 0.14010$ seconds. But, to guarantee the settlement of transient behavior, the FW-H solver starts after 20×10^3 time steps and runs for an extra 8.0×10^3 time steps to obtain acoustic data. The noise spectra of corresponding simulations are generated by a total of 3960 samples. Sample number remained constant for the summation and calculation of overall PSD noise level. It is noteworthy that although noise-generating flow is generally turbulent, unsteady, and nonlinear, usually linear sound field is radiated. Generally, acoustic pressure at far-field is below atmospheric pressure which proves the validity of the superposition principle [72]. Therefore, LAGOON NLG overall noise is achieved as the incoherent sum of the calculated noise from each cross-section.

8.4 Results and Discussion

All acoustic results given here are obtained according to the multiple 2D cross-sections of 2-NLG (LAGOON). First, receiver location effect is plotted taking into account three different SCL values to evaluate the effect of this factor on the prediction of noise. Secondly, the importance of including quadrupole source is considered by specifying different porous surfaces (PorSurfs) in the vicinity of the bluff body. Then, the focus is dedicated to the lower part of the LG due to its complexity. Finally, the overall noise resulting from each cross-section is obtained and then validated using 3D numerical and experimental results.

8.4.1 The Effects of the Receiver Locations

One of the important parameters that impacts the prediction accuracy of far-field noise is the location of the receiver with respect to each 2D cross-section. To examine this parameter, three different microphone positions are chosen. These microphones are located at the same distance of 6 m perpendicular to the centerline of the corresponding cross-section but at different angles. The three receivers are positioned orthogonally apart from each other (i.e. 90°). Figure 8.8a–c shows the SPL results at 2DMic (X-axis), 2DMic (Y-axis), and 3DMic (Z-axis), respectively, for Sec.1, for example. It can be noted that both 2DMic (X-axis) and 2DMic (Y-axis) have similar SPL results. For instance, the SCL influences the predicted noise spectrum over the entire frequency band (100 Hz–10 kHz). In other words, it is noted that the SPL noise levels with different SCL amplitudes are in a constant offset from each other. However, when the receiver is placed along the span of the cylinder (3DMic), different results are observed. The constant offset pattern of the noise spectra is only seen at the low-frequency range (100–500 Hz), whereas the noise level collapses at the high-frequency range ($f \geq 1$ kHz). In general, for both 2DMic locations, it is noted that doubling the SCL results in a doubling of the acoustic energy (i.e. $\approx +3$ dB).

Figure 8.9a–c represents the time history of the corresponding acoustic pressure, p', data of the noise spectra shown in Figure 8.8a–c, respectively. Interestingly, the recorded p' at the 2DMic (Y-axis) resembles the same oscillation pattern of the lift coefficient, C_l, as can be seen in Figure 8.10. This implies that the C_l of near-field flow perturbation has the greatest influence in exciting the medium (i.e. surrounding air molecules at rest). However, the drag coefficient, C_d, is the secondary near-field flow fluctuation.

The former is known as von Karman vortex street which is associated with the flow past a cylinder bluff body which exhibits a typical transversal oscillation of its wake (i.e. lateral flow instability that is perpendicular to the direction of flow). The latter is associated with drag force that is parallel to the flow direction and only exerts pressure on the back side of the cylinder.

The previous observation of the near-field flow fluctuation results in local pressure variation with respect to the atmospheric pressure, in which the sound is generated and then the acoustic wave (sound of speed, $c_o = 340$ m/s) is propagated toward the far-field region. Thus, the noise generation is only considered as a by-product of the unsteady flow structure. As mentioned, the wake's oscillation is connected with the oscillation of the C_l around the upper and lower sides of the cylinder. For the compact source ($L_a \ll \lambda$), a dipole source is

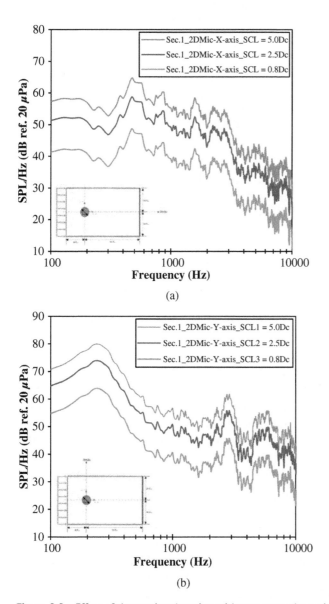

(a)

(b)

Figure 8.8 Effect of the receiver location with respect to the variation of the SCL for Sec.1 of the LAGOON NLG at (a) 2DMic (X-axis), (b) 2DMic (Y-axis), and (c) 3DMic (Z-axis).

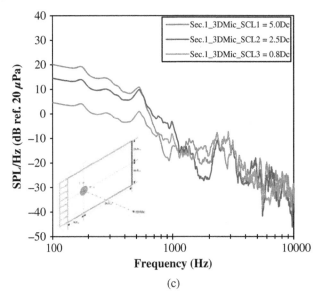

(c)

Figure 8.8 (*Continued*)

detected at the far-field noise spectrum. Figure 8.11 illustrates the influence of the receiver locations and the SCL on the predicted noise.

When the SCL is short (Figure 8.11b), the source is compact and the directivity of the generated sound is a dipole, with its lobes pointing in the direction of the lift. In contrast, for the larger SCL (Figure 8.11c), the compactness condition is violated, and the directivity becomes flatter in the lateral direction and expands toward the X-axis (Figure 8.11e). In fact, this is also observed in the acoustic pressure data that is recorded at the 3DMic, as shown in Figure 8.9c. Therefore, the SCL should be assumed carefully in accordance with the compactness condition to ensure that the primary noise is extracted. In addition, by assuming the short SCL, the near-field flow would still preserve its coherence along the spanwise direction.

8.4.2 The Effects of the Acoustic Source

Another important parameter is the location of the acoustic sources. When the solid surface of each cross-section is specified as the source of acoustic, only near-field unsteady surface pressure is considered. Consequently, only the contribution from the dipole sources is included. For instance, Figure 8.12a depicts the signals of the far-field noise for Sec.4 at three different locations such as receiver-1, receiver-2 at the 2DMic Y-axis at −6 m and at +6 m, respectively, and receiver-3 at the 3DMic. It is observed that, the PSD spectrum changes in accordance with the recorded far-field acoustic pressure data. Different amplitude and frequency contents are captured accordingly. Interestingly, the results of the two opposite 2DMic receivers indicate that the wall pressure distribution around Sec.4 is not symmetrical. This indicates that at some frequencies the peaks are not aligned but instead they have different patterns in terms of amplitude level or phase shifting. To consider the effect of

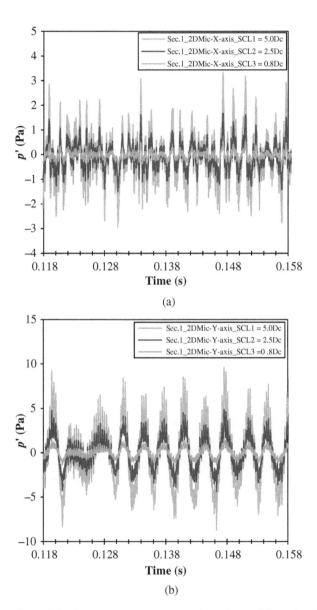

Figure 8.9 Acoustic pressure results with different SCL and at different receiver locations for Sec.1 of the LAGOON NLG at (a) 2DMic (X-axis), (b) 2DMic (Y-axis), and (c) 3DMic (Z-axis).

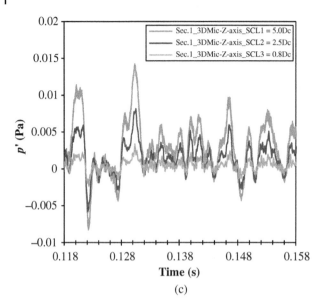

Figure 8.9 (*Continued*)

other sources such as quadrupole, some of the permeable surfaces are created away from Sec.4, as shown in the bottom right corner of Figure 8.12b. Three distinct porous surfaces are named PorSurf-1, PorSurf-2, and PorSurf-3. The effect of considering the contributions from different combinations of acoustic sources is investigated. Figure 8.12b illustrates the significance of accounting for different acoustic sources generated at different regions in the near-field domain of Sec.4. The same behaviors are noticed for the other cross-sections.

8.4.3   The LAGOON NLG Overall Far-Field Acoustic Results

All the obtained results for the primary five cross-sections at 3DMic are summed up and compared with experimental 3D numerical data. It is noted that the total PSD noise level is more accurate than before, where the tonal peak at 1.5 kHz is slightly evident, as shown in Figure 8.13a. Moreover, a better matching in both low and high frequency regions is observed. However, its amplitude and phase are still not perfectly matching the experimental tone. Thus, more investigations are carried out to add the results of the other cases. The PSD signal for Sec.4-C at the 3DMic is then added to the previous result. Figure 8.13b shows the signal of the overall PSD according to the summation of the signals of the primary five cross-sections and Sec.4-C. Based on this, a better result is obtained in terms of the peak amplitude and phase at 1.5 kHz compared with Figure 8.13a.

Nevertheless, the amplitude is still smaller than the experiment. The same procedure is repeated for the rest of the cases but there is not much improvement. Therefore, the signal based on this combination of acoustic sources collected at the 3DMic is taken into account. Then, the signal of Sec.4 based on the combination of solid and PorSurf-1 and PorSurf-2 is added to the previous primary five cross-sections as well as Sec.4-C. Figure 8.14 presents the corresponding PSD spectra of all these cross-sections and the total PSD noise level based on

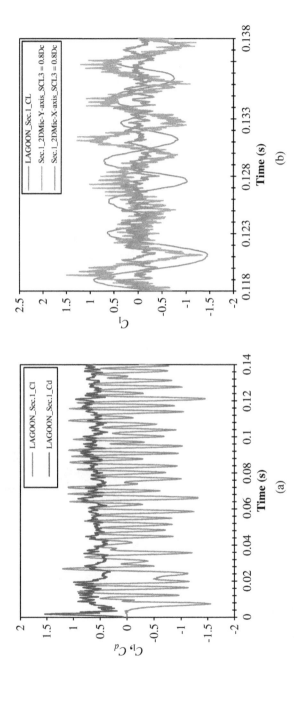

Figure 8.10 Near-field flow fluctuations for Sec.1 of the LAGOON NLG: (a) time history for the lift and drag coefficients; and (b) comparison with the acoustic pressure at 2DMic.

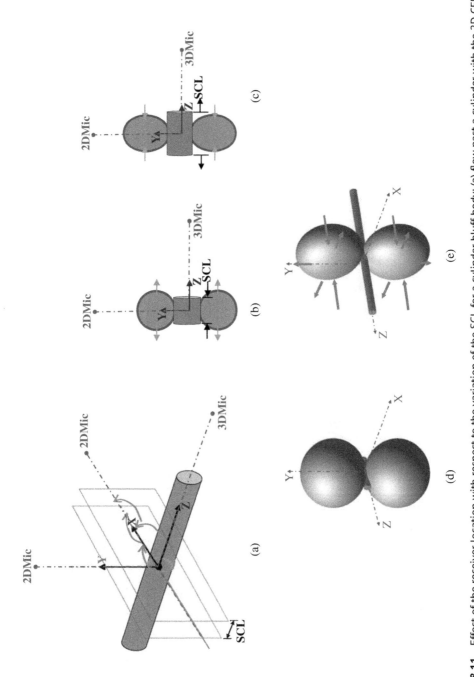

Figure 8.11 Effect of the receiver location with respect to the variation of the SCL for a cylinder bluff body: (a) flow past a cylinder with the 2D CFD mapping along the span using the SCL and receiver locations; (b) and (c) illustrate the result of the dipole-like source with respect to short and long SCL, respectively; and (d) and (e) 3D visualization of the dipole source.

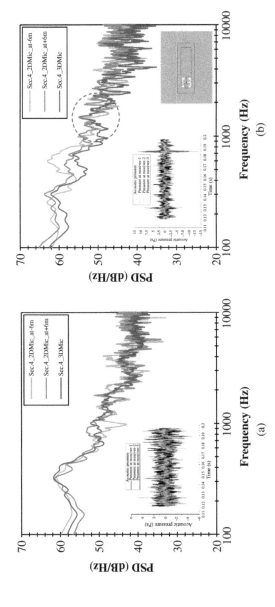

Figure 8.12 PSD and acoustic pressure signals of Sec.4 at three different receiver locations: (a) solid surface alone; and (b) combination of solid, PorSurf-1 and PorSurf-2 acoustic sources.

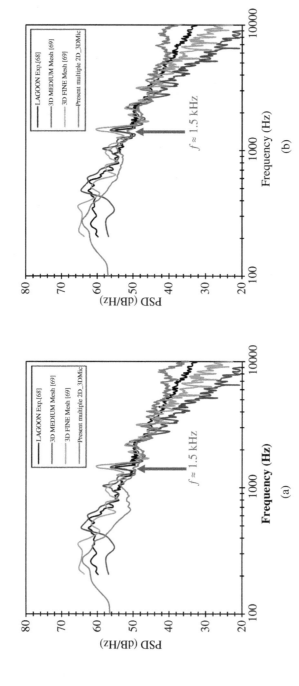

Figure 8.13 PSD signals at 3DMic for (a) the primary five cross-sections only and (b) the primary five cross-section signals including Sec.4-C and comparison with experimental and 3D numerical data. Source: [66]. Licensed under CC BY 4.0.

Table 8.2 Quantitative comparison between the experimental, 3D numerical, and present multiple 2D results for the LAGOON NLG.

Case	Tonal peak frequency (kHz)	Error (%)[a]
LAGOON exp. [68]	1.50	—
3D FINE mesh [69]	1.53	1.96
3D MEDIUM mesh [69]	1.52	1.32
Present multiple 2D_3DMic	1.51	0.66

a) Error percentage $= \left| \frac{f_{Num.} - f_{Exp.}}{f_{Exp.}} \right| \times 100$.

their summations. It can be clearly seen that a much better agreement with experimental data is obtained. Tonal peak at 1.5 kHz is successfully reproduced with much better amplitude level compared with the previous summations. Also, a good matching is achieved at both the low- and high-frequency regions, as highlighted in Figure 8.14b.

Table 8.2 compares the captured tonal peak between the experimental, 3D numerical, and the present multiple 2D results. It can be noted that the present multiple 2D result is accurate with only 0.66% error relative to the experimental tonal peak, f_{Exp}. This shows the capability of the proposed method in capturing the essential peaks.

Each simulation requires about 10–16 hours (i.e. two to three days were required to perform all five 2D simulations) to achieve acoustic data. Computations were conducted using a single core on a 64-bit, 3.60 GHz Intel® Core™ i7-4790 with 32 GB memory. However, the 3D simulation [69] with the fine mesh (15×10^6 cells) was run on parallel mode using supercomputers (with 600 cores) provided by Airbus France. This 3D simulation took about 109 hours (i.e. about five days) to obtain the results of only $t = 0.12$ seconds (e.g. $\Delta t = 3.0 \times 10^{-7}$ seconds). It can be clearly noted that our proposed method has significantly less computational cost than 3D simulation.

8.5 Summary and Conclusions

In this chapter, the development of a novel far-field noise prediction methodology and its numerical assessments for 2-NLG (LAGOON) was presented. The main reason for developing such a model is so that it can be used as a valuable framework for obtaining a quick estimate of LG far-field noise. In the proposed multiple two-dimensional simulations method (M2DSM), a number of issues within the current 3D LG simulations are addressed, including quadrupole source contribution, high number of cells (e.g. more than 10 million cells), long computational time (e.g. more than several months), CPU time and memory consumption (i.e. supercomputers are required). The development of the M2DSM was shown to successfully avoid all of these dilemmas by decomposing or breaking down the full 3D LG model into multiple 2D domains.

The 3D LAGOON NLG model was divided into multiple 2D cross-sections to take into account various configurations. In total, five different cross-sections locations were adopted

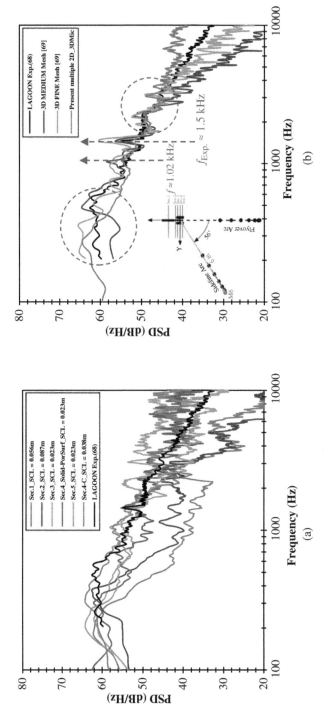

Figure 8.14 Signals at 3DMic for (a) all the cross-section signals, where Sec.4 is based on the combination of solid, PorSurf-1 and PorSurf-2 acoustic sources and (b) overall PSD and comparison with experimental and 3D numerical data. Source: [66]. Licensed under CC BY 4.0.

according to sizes, geometric complexities, and the number of components. To compensate for missing CFD data along the span, three different SCL values were assumed according to a perfectly correlated flow. Overall LG noise was obtained as the incoherent sum of calculated noise obtained from all cross-sections. It was noted that the SCL has a great influence on the calculated level of noise over the entire frequency spectrum. It was also observed that the receiver location with respect to the flow direction alters both the frequency and amplitude of far-field noise spectra. The results based on the location of 3DMic revealed better matching with experimental results.

After that, focus was directed to conduct extensive investigations on the effect of the receiver and acoustic source locations on predicted far-field noise. It was discovered that both receiver and source locations have a great impact on the level and details of the noise spectrum. In fact, a better agreement was obtained when quadrupole source effect on the proximity of Sec.4 was included. This confirms that the inclusion of the quadrupole source is significant, where it was shown that the accuracy of the predicted far-field noise was improved. By using the proposed M2DSM, the issue of the computational cost of the quadrupole term is no longer an issue, where the FW-H porous surface method can be easily implemented. The total PSD noise level at the 3DMic for the summation of the primary five cross-sections, Sec.4-C, and Sec.4 with solid, PorSurf-1, and PorSurf-2 as acoustic sources complied well with experimental and 3D numerical data. Tonal peak at 1.5 kHz was successfully reproduced and complies well with experimental tone with only 0.66% error. This reveals the capability of M2DSM in reasonably obtaining 3D LG aerodynamic noise at low computational cost. It was shown that all the 2D simulations took less than two to three days to obtain the acoustic data using only a single-core processor. The calculations were performed using a single core on a 64-bit, 3.60 GHz Intel Core™ i7-4790 with 32 GB memory. Thus, it could be utilized as a useful framework and a quick design tool to evaluate the noise emitted from each LG component, so that new low-noise LG designs can be proposed and investigated.

References

1 Busquin, P., Argüelles, P., Bischoff, M. et al. (2001). European aeronautics: a vision for 2020 – a synopsis. *Air & Space Europe* 3: 16–18.

2 Chow, L., Mau, K., and Remy, H. (2002). Landing gears and high lift devices airframe noise research. In: *8th AIAA/CEAS Aeroacoustics Conference & Exhibit*, Breckenridge, CO, USA (17–19 June 2002), 2408. AIAA.

3 Grosche, F., Schneider, G., Stiewitt, H. et al. (1997). Wind tunnel experiments on airframe noise sources of transport aircraft. In: *3rd AIAA/CEAS Aeroacoustics Conference*, Atlanta, GA, USA (12–14 May 1997), 1642. AIAA.

4 Dobrzynski, W. (2010). Almost 40 years of airframe noise research: what did we achieve? *J. Aircr.* 47: 353–367.

5 Brice, S. (2012). Ties that bind: FAA enforcement of grant assurances-the Santa Monica airport case. *The Air and Space Lawyer* 25: 9.

6 Argüelles, P., Bischoff, M., Busquin, P. et al. (2001). European Aeronautics: A Vision for 2020. Advisory Council for Aeronautics Research in Europe.

7 European Commission (2011). Flightpath 2050: Europe's Vision for Aviation. Report of the High Level Group on Aviation Research.

8 Delfs, J., Bertsch, L., Zellmann, C. et al. (2018). Aircraft noise assessment—from single components to large scenarios. *Energies* 11: 429.

9 Choudhari, M. and Yamamoto, K. (2012). Integrating CFD, CAA, and experiments towards benchmark datasets for airframe noise problems. 5th Symposium on Integrating CFD and Experiments in Aerodynamics (Integration 2012), Tokyo.

10 Lopes, L.V., Redonnet, S., Imamura, T. et al. (2015). Variability in the propagation phase of CFD-based noise prediction: Summary of results from Category 8 of the BANC-III Workshop. In: *21st AIAA/CEAS Aeroacoustics Conference*, Dallas, TX, USA (22–26 June 2015), 2845. AIAA.

11 Choudhari, M. and Lockard, D. (2016). Simulations & Measurements of Airframe Noise: a BANC Workshops Perspective.

12 Singer, B.A. and Guo, Y. (2004). Development of computational aeroacoustics tools for airframe noise calculations. *Int. J. Comput. Fluid Dyn.* 18: 455–469.

13 Farassat, F. and Casper, J. (2006). Towards an airframe noise prediction methodology: Survey of current approaches. In: *44th AIAA Aerospace Sciences Meeting and Exhibit*, Reno, NV, USA (9–12 January 2006), 210. AIAA.

14 Schlottke-Lakemper, M., Yu, H., Berger, S. et al. (2017). A fully coupled hybrid computational aeroacoustics method on hierarchical Cartesian meshes. *Comput. Fluids* 144: 137–153.

15 Lazos, B. (2002). Surface topology on the wheels of a generic four-wheel landing gear. *AIAA J.* 40: 2402–2411.

16 Dobrzynski, W. and Buchholz, H. (1997). Full-scale noise testing on Airbus landing gears in the German Dutch wind tunnel. In: *3rd AIAA/CEAS Aeroacoustics Conference*, Atlanta, GA, USA (12–14 May 1997), 1597. AIAA.

17 Heller, H.H. and Dobrzynski, W.M. (1977). Sound radiation from aircraft wheel-well/landing-gear configurations. *J. Aircr.* 14: 768–774.

18 Ravetta, P., Burdisso, R., and Ng, W. (2004). Wind tunnel aeroacoustic measurements of a 26%-scale 777 main landing gear. 10th AIAA/CEAS Aeroacoustics Conference, Manchester.

19 Guo, Y., Yamamoto, K.J., and Stoker, R.W. (2006). Experimental study on aircraft landing gear noise. *J. Aircr.* 43: 306–317.

20 Vuillot, F., Lupoglazoff, N., Luquent, D. et al. ((2012). Hybrid CAA solutions for nose landing gear noise. In: *18th AIAA/CEAS Aeroacoustics Conference (33rd AIAA Aeroacoustics Conference)*, Colorado Springs, CO, USA (4–6 June 2012), 2283. AIAA.

21 Seror, C., Sagaut, P., and Bélanger, A. (2004). A numerical aeroacoustic analysis of a detailed landing gear. 10th AIAA/CEAS Aeroacoustics Conference, Manchester.

22 Dobrzynski, W., Chow, L., Guion, P., and Shiells, D. (2000). A European study on landing gear airframe noise sources. In: *6th Aeroacoustics Conference and Exhibit*, Lahaina, HI, USA (12–14 June 2000), 1971. AIAA.

23 Dobrzynski, W., Chow, L., Guion, P., and Shiells, D. (2002). Research into landing gear airframe noise reduction. In: *8th AIAA/CEAS Aeroacoustics Conference & Exhibit*, Breckenridge, CO, USA (17–19 June 2002), 2409. AIAA.

24 Fink, M.R. (1979). Airframe noise prediction method. *Tech. Rep. FAA-RD-77-29.*

25 Guo, Y., Yamamoto, K., and Stoker, R. (2004).An empirical model for landing gear noise prediction. 10th AIAA/CEAS Aeroacoustics Conference, Manchester.

26 Smith, M. and Chow, L. (1998). Prediction method for aerodynamic noise from aircraft landing gear. In: *4th AIAA/CEAS Aeroacoustics Conference*, Toulouse, France (2–4 June 1998), 2228. AIAA.

27 Casalino, D., Diozzi, F., Sannino, R., and Paonessa, A. (2008). Aircraft noise reduction technologies: a bibliographic review. *Aerosp. Sci. Technol.* 12: 1–17.

28 Dobrzynski, W., Ewert, R., Pott-Pollenske, M. et al. (2008). Research at DLR towards airframe noise prediction and reduction. *Aerosp. Sci. Technol.* 12: 80–90.

29 Dobrzynski, W., Chow, L.C., Smith, M. et al. (2010). Experimental assessment of low noise landing gear component design. *Int. J. Aeroacoust.* 9: 763–786.

30 Thomas, F., Kozlov, A., and Corke, T. (2005). Plasma actuators for landing gear noise reduction. 11th AIAA/CEAS Aeroacoustics Conference, Monterey, CA.

31 Wu, G., Lu, Z., Xu, X. et al. (2019). Numerical investigation of aeroacoustics damping performance of a Helmholtz resonator: effects of geometry, grazing and bias flow. *Aerosp. Sci. Technol.* 86: 191–203.

32 Belyaev, I., Zaytsev, M.Y., Kopiev, V., and Pankratov, I. (2019). Experimental research on noise reduction for realistic landing gear geometries. *Acoust. Phys.* 65: 297–306.

33 Curle, N. (1955). The influence of solid boundaries upon aerodynamic sound. *Proc. R. Soc. London, Ser. A* 231: 505–514.

34 Golub, R.A., Sen, R., Hardy, B. et al. (2004). Airframe noise sub-component definition and model. *Tech. Rep. NASA/CR-2004-213255*.

35 Guo, Y. (2005). A statistical model for landing gear noise prediction. *J. Sound Vib.* 282: 61–87.

36 Lopes, L., Brentner, K., Morris, P. et al. (2005). Complex landing gear noise prediction using a simple toolkit. 43rd AIAA Aerospace Sciences Meeting and Exhibit, Reno, NV.

37 Tam, C.K. (1995). Computational aeroacoustics-issues and methods. *AIAA J.* 33: 1788–1796.

38 Roe, P. (1992). Technical prospects for computational aeroacoustics. Aeroacoustics Conference, Aachen

39 Long, S.L., Nie, H., and Xu, X. (2012). Aeroacoustic study on a simplified nose landing gear. *Appl. Mech. Mater.* 184: 18–23.

40 Crighton, D., Dowling, A.P., Williams, J.F. et al. (2012). *Modern Methods in Analytical Acoustics: Lecture Notes*. Springer Science & Business Media.

41 Lighthill, M.J. (1952). On sound generated aerodynamically I. General theory. *Proc. R. Soc. London, Ser. A* 211: 564–587.

42 Ffowcs Williams, J.E. and Hawkings, D.L. (1969). Sound generation by turbulence and surfaces in arbitrary motion. *Philos. Trans. R. Soc. London, Ser. A* 264: 321–342.

43 Lyrintzis, A.S. (1994). The use of Kirchhoff's method in computational aeroacoustics. *J. Fluids Eng.* 116 (4): 665–676.

44 Brentner, K.S. and Farassat, F. (1998). Analytical comparison of the acoustic analogy and Kirchhoff formulation for moving surfaces. *AIAA J.* 36: 1379–1386.

45 van Mierlo, K., Takeda, K., and Peers, E. (2010). Computational analysis of the effect of bogie inclination angle on landing gear noise. 16th AIAA/CEAS Aeroacoustics Conference, Stockholm.

46 Lockard, D., Khorrami, M., and Li, F. (2004). High resolution calculation of a simplified landing gear. 10th AIAA/CEAS Aeroacoustics Conference, Manchester.

47 Spalart, P.R., Shur, M.L., Strelets, M.K., and Travin, A.K. (2011). Initial noise predictions for rudimentary landing gear. *J. Sound Vib.* 330: 4180–4195.

48 Sanders, L., Manoha, E., Khelil, S., and Francois, C. (2011). LAGOON: CFD/CAA coupling for landing gear noise and comparison with experimental database. 17th AIAA/CEAS Aeroacoustics Conference, Portland, OR.

49 Souliez, F.J., Long, L.N., Morris, P.J., and Sharma, A. (2002). Landing gear aerodynamic noise prediction using unstructured grids. *Int. J. Aeroacoust.* 1 (2): 115–135.

50 Vatsa, V.N., Khorrami, M.R., Rhoads, J., and Lockard, D.P. (2015). Aeroacoustic simulations of a nose landing gear using FUN3D on Pointwise unstructured grids. 21st AIAA/CEAS Aeronautics Conference, Dallas, TX.

51 Vatsa, V.N., Khorrami, M.R., and Lockard, D.P. (2017). Aeroacoustic simulations of a nose landing gear with FUN3D: A grid refinement study. 23rd AIAA/CEAS Aeroacoustics Conference, Denver, CO.

52 de Abreu, R.V., Jansson, N., and Hoffman, J. (2016). Computation of aeroacoustic sources for a gulfstream G550 nose landing gear model using adaptive FEM. *Comput. Fluids* 124: 136–146.

53 Liu, W., Kim, J.W., Zhang, X. et al. (2013). Landing-gear noise prediction using high-order finite difference schemes. *J. Sound Vib.* 332: 3517–3534.

54 Wagner, C., Hüttl, T., and Sagaut, P. (2007). *Large-Eddy Simulation for Acoustics*. Cambridge University Press.

55 Farassat, F. and Brentner, K.S. (1998). Supersonic quadrupole noise theory for high-speed helicopter rotors. *J. Sound Vib.* 218: 481–500.

56 Farassat, F. (2007). Derivation of Formulations 1 and 1A of Farassat. *Tech. Rep. NASA/TM-2007-214853.*

57 Wang, M., Freund, J.B., and Lele, S.K. (2006). Computational prediction of flow-generated sound. *Annu. Rev. Fluid Mech.* 38: 483–512.

58 Di Francescantonio, P. (1997). A new boundary integral formulation for the prediction of sound radiation. *J. Sound Vib.* 202: 491–509.

59 Choi, W., Choi, Y., Hong, S. et al. (2016). Turbulence-induced noise of a submerged cylinder using a permeable FW–H method. *Int. J. Nav. Archit. Ocean Eng.* 8: 235–242.

60 Manoha, E., Bulté, J., and Caruelle, B. (2008). LAGOON: an experimental database for the validation of CFD/CAA methods for landing gear noise prediction. 14th AIAA/CEAS Aeroacoustics Conference (29th AIAA Aeroacoustics Conference), Vancouver, BC.

61 Szepessy, S. (1994). On the spanwise correlation of vortex shedding from a circular cylinder at high subcritical Reynolds number. *Phys. Fluids* 6: 2406–2416.

62 De la Puente, F.; Sanders, L.; and Vuillot, F. (2014). On LAGOON nose landing gear CFD/CAA computation over unstructured mesh using a ZDES approach. 20th AIAA/CEAS Aeroacoustics Conference, Atlanta, GA.

63 Rodarte Ricciardi, T.; Azevedo, P.; Wolf, W.; and Speth, R. (2017). Noise prediction of the LAGOON landing gear using detached Eddy simulation and acoustic analogy. 23rd AIAA/CEAS Aeroacoustics Conference, Denver, CO.

64 Heller, H.H. and Dobrzynski, W.M. (1977). Sound radiation from aircraft wheel-WelVLanding-gear configurations. *J. Aircr.* 14: 768–774.

65 Demartino, C. and Ricciardelli, F. (2017). Aerodynamics of nominally circular cylinders: a review of experimental results for civil engineering applications. *Eng. Struct.* 137: 76–114.

66 Alqash, S., Dhote, S., and Behdinan, K. (2019). Predicting far-field noise generated by a landing gear using multiple two-dimensional simulations. *Appl. Sci.* 9 (21): 4485.

67 Choudhari, M. and Lockard, D. (2016). Simulations and measurements of airframe noise: A BANC workshops perspective. In: *Proceedings of NATO STO-MP-AVT-246 Specialists Meeting on Progress and Challenges in Validation Testing for Computational Fluid Dynamics*, Zarazoga, Spain (26–28 September 2016).

68 Manoha, E., Bulté, J., Ciobaca, V., and Caruelle, B. (2009).LAGOON: further analysis of aerodynamic experiments and early aeroacoustics results. 15th AIAA/CEAS Aeroacoustics Conference (30th AIAA Aeroacoustics Conference), Miami, FL.

69 Giret, J.C., Sengissen, A., Moreau, S., and Jouhaud, J.C. (2013). Prediction of LAGOON landing-gear noise using an unstructured LES solver. In: *Proceedings of the 19th AIAA/CEAS Aeroacoustics Conference*, Berlin, Germany (27–29 May 2013), 2113. AIAA.

70 Roshko, A. (1961). Experiments on the flow past a circular cylinder at very high Reynolds number. *J. Fluid Mech.* 10: 345–356.

71 Norberg, C. (2003). Fluctuating lift on a circular cylinder: review and new measurements. *J. Fluids Struct.* 17: 57–96.

72 Lopes, L.V. (2009). A new approach to complete aircraft landing gear noise prediction. PhD thesis. The Pennsylvania State University.

9

Vibration Transfer Path Analysis of Aeroengines Using Bond Graph Theory

Seyed Ehsan Mir-Haidari and Kamran Behdinan

Department of Mechanical & Industrial Engineering, Advanced Research Laboratory for Multifunctional Lightweight Structures (ARL-MLS), University of Toronto, Toronto, Canada

9.1 Introduction

Low profit margins faced by the airline operator have forced the sector to seek the development of efficient aeroengines. The demand for more efficient aeroengines has been amplified with worldwide increase in prices of energy commodities. In order to meet this demand, aeroengine manufacturers have focused on developing lightweight aeroengines that use advanced lightweight materials with higher power performance and output. By developing lightweight aeroengines, the structural response and sensitivity to internal excitation loadings attributed to rotor system unbalance forces caused by mass eccentricity is amplified and increased [1]. The rotor system mass eccentricity which leads to unbalance loads is primarily caused by limitations in the manufacturing process of the rotor system [1]. Large unbalance forces originating from the aeroengine makes the system an active contributor of noise and vibration transfer to the aircraft fuselage. The propagation of vibration energy from the aeroengine to the fuselage significantly affects the wellbeing and comfort of the passengers on board. Minimizing the transfer noise and vibration in the aircraft has gained renowned interest by researchers, seeking advanced active and passive methodologies to minimize vibration transfer in the aircraft by implementing various transfer path analysis (TPA) methods [2–11].

TPA is a combination of experimental and theoretical based methodologies intended to analyze vibration energy propagation throughout a system [12]. TPA is a trustworthy and important methodology which is used to study vibration energy propagation throughout the systems to propose guidelines in minimizing such energy transfer in engineering systems in various industry sectors [13]. TPA found its way into mechanical engineering a hundred years ago by adaptation of electric network analogies [12]. Road-induced noise in the cabin of vehicles has been widely used as the prime experimental benchmark in various publications for TPA. The need for TPA has become paramount due to increased demand in minimizing unwanted vibrations, leading to improved safety and comfort [12].

In this chapter, the bond graph TPA method is proposed and fully explained. The proposed methodology combined with the transmissibility principle is used to perform a comprehensive TPA analysis on a reduced aeroengine model. Using this approach, transmissibility between various aeroengine components is determined to identify critical operational frequencies and vibration energy propagation pathways of a reduced aeroengine representation. Using this approach, vibration energy propagation can be minimized. The bond graph TPA approach allows implementation of cumbersome design modifications prior to any manufacturing; this reduces overall development costs, which is one of the main goals of the aerospace industry sector [14, 15]. Using bond graph methodology, passive vibration control guidelines are proposed that would include aeroengine structural modifications to minimize the overall vibration energy propagation in a commercial aircraft aeroengine.

9.2 Overview of TPA Methodologies

TPA is concerned with an active vibrating component(s) (i.e. engines, rotors, gearing systems, etc.) and the transfer of vibration energy from the source to the final receiving location of interest on the structure through various structural connections. A TPA system structure is constituted of active and passive systems which are connected by various mechanical systems. TPA methodology permits differentiation of the active and passive characteristics of the structure under analysis. This approach allows the determination of possible path contributions from the source to the receiver of the system under analysis, which would allow the determination of dominant vibration transfer paths [12]. Figure 9.1 illustrates the active component (excitation source) of the system that generates the excitation forces $F_i(\omega)$, and the passive receiving part that receives these forces. H_{ij} represents the components of the transfer function (TF) between the input and output of the system.

The main three components for a TPA analysis are as follows:

- **Active component:** The active components of the system can be structural. For example, an excitation source in an aircraft includes the aeroengine or the landing gears.

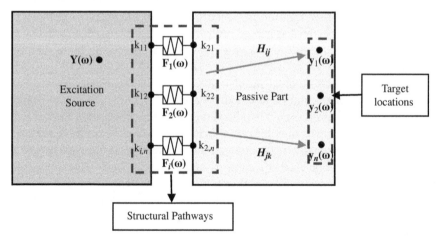

Figure 9.1 Structural vibration transfer path schematic.

- **Target locations (passive part):** These can be vibrations perceived by the user, such as the vibrations of the fuselage in an aircraft.
- **Transfer paths (mechanical connections):** Mechanical connections where vibration energy is propagated starting from the active component to the receiving point on the structure. Examples include bolted joints in aeroengine casing assemblies.

To perform a conventional TPA, it is first necessary to identify the operational loads during transient conditions (i.e. run-up or run-down conditions). Secondly, system frequency response functions (FRFs) must be estimated between the excitation source and the target location. It should be noted that measuring the FRF between the excitation source and the target location is not an accuracy factor in performing TPA. However, the main accuracy factor that contributes to discrepancies is the identification of operational loads [16]. The presently available techniques only allow an estimation of operational loads, such as requiring a prior knowledge of mount stiffnesses or removal of the active component [13]. Three possible techniques are currently available to measure the operational loads for classical TPA; these have been extensively studied in various publications [3–6, 17, 18].

9.2.1 Measuring Interface Loads Using the Classical TPA Approach

Using this method, the operational loads acting on the structure are directly measured. This is achieved by placing load cell sensors between the active excitation source and the passive structure. This approach has the disadvantage of introducing overall changes to the system characteristics and being complicated to perform. Two methods are proposed in the literature that try to overcome the challenges faced by this method.

9.2.1.1 Mount Stiffness Measurement Technique

Using the mount stiffness measurement (MSM) technique, the load is measured using the measured displacement between the source and the passive receiving component using the following equation [17].

$$F_p(\omega) = K_p(\omega) \cdot (X_{ap}(\omega)X_{pp}(\omega)) = \frac{K_p(\omega) \cdot (A_{ap}(\omega)A_{pp}(\omega))}{-\omega^2} \tag{9.1}$$

where $X_{ap}(\omega)$ and $A_{ap}(\omega)$ are the mount displacement and acceleration at the active side, respectively. $X_{pp}(\omega)$ and $A_{pp}(\omega)$ are the mount displacement and acceleration at the passive side, respectively. $K_p(\omega)$ is the measured stiffness value and $F_p(\omega)$ is the operational force at path p.

There are several notable disadvantages in implementing the MSM technique. First, determination of torsional stiffness causes some difficulties. In addition, the MSM technique requires the measurement of all FRFs for all possible pathways in the structure; therefore, all active sources would need to be uncoupled from the structure to identify them, which introduces some physical measurement challenges.

One of the main advantages associated with this method is that any measurement error that occurs for a path does not contribute to error stack up of other possible paths in the system. The MSM technique is best to be implemented in the lower frequency range. This is attributed to the fact that this model illustrates amplitude dependent nonlinearities at higher frequencies, which directly affects the overall accuracy of the model [12, 17].

9.2.1.2 Matrix Inversion Method

Using this method, the required loads are obtained from operational measurements at indicator points close to the active vibration source. The formulation is given below, which is based on system matrix inversion [17]:

$$(I_i(\omega))_{Ni} = \left[H^{\frac{I}{F}}_{i,j}(\omega) \, H^{\frac{I}{Q}}_{i,j}(\omega) \right]_{Ni,Nj} \cdot \left\{ \begin{array}{c} F_j(\omega) \\ Q_j(\omega) \end{array} \right\}_{Nj} \tag{9.2}$$

where $I_i(\omega)$ are the operational data at indicator i. $H^{\frac{I}{F}}_{i,j}(\omega)$ is the TF between forces at j and responses at location i and $H^{\frac{I}{Q}}_{i,p}(\omega)$ is the TF for volume acceleration at path j and responses at location i. Ni represents the total number of indicators in the system, $F_p(\omega)$ is force at j, and $Q_p(\omega)$ is the acceleration at j.

The loads are determined by inversing the provided matrix in Eq. (9.2). For numerical accuracy, the pseudo inversion method is implemented:

$$\left\{ \begin{array}{c} F_j(\omega) \\ Q_j(\omega) \end{array} \right\}_{Nj} = \left[H^{\frac{I}{F}}_{j,j}(\omega) \, H^{\frac{I}{Q}}_{j,j}(\omega) \right]^{+}_{Nj,Nj} \cdot \left\{ \begin{array}{c} A_j(\omega) \\ P_j(\omega) \end{array} \right\}_{Nj} \tag{9.3}$$

where $A_p(\omega)$ is the acceleration at j and $P_j(\omega)$ is the operational pressure near the structure at path location j.

Using this method, the loads are determined without any information regarding the connection of the source to the system. The disadvantages associated with this method are:

- Any load in the system should be represented in the theoretical model.
- If a pathway is not included in the model, erroneous results will be obtained for all other pathways.
- The approach is not computationally efficient.

9.2.2 Operational Path Analysis

To bypass the difficulties faced by classical TPA, many other TPA methods have been studied and proposed in the literature; all of them propose various different trade-offs [13]. One of the proposed techniques studied in the literature is operational path analysis (OPA). This TPA approach is based on the idea of the transmissibility principle that has been around since the 1990s [19, 20]. By implementing the OPA approach, only operational data are measured compared with the approach of the classical TPA [13], such as the FRF measurement.

All measurements are performed on the operating system, meaning that only the input excitations are utilized to determine various propagation paths in the system. In comparison with the classical TPA which is based on FRF methodology, it is important to emphasize the following [4]:

- The method is very time efficient to perform.
- Since operational data are measured in this method, essentially the operational influences are automatically included and accounted for.

The OPA technique determines the linearized TF matrix between the selected input and output. Furthermore, OPA calculations are transmissibility based instead of TFs, making them dependent on system loading conditions.

9.2.2.1 OPA Theory

To develop the OPA theory, an arbitrary linearized system is expressed as:

$$y(\omega) = H(\omega) \cdot x(\omega) \tag{9.4}$$

where x represents the input to the system defined by $x = [u_x f_x p_x]$, y is the system response at selected target locations defined by $y = [u_y f_y p_y]$, and $H(\omega)$ is the structure-borne TF between the input and the output. As a rule of thumb, the size of the external excitation vector x should be twice the size of the response vector y to avoid numerical difficulties in calculations. The TF is defined as:

$$H_{ij} = \left. \frac{y_i}{x_j} \right|_{x_k=0} ; k \neq j \tag{9.5}$$

The TF matrix is typically obtained using experimental means by using an external excitation as the only degree of freedom (DOF) input, and the output is selected as the response of the system. In practical applications, this approach is difficult to implement since the inputs to the system might not only be mechanical, but acoustical as well, which makes this method very challenging to perform and requiring a huge investment in time and resources.

To address this challenge, the OPA determines all the variables of the TF in a single measurement when all excitations are simultaneously present. To achieve this, the transpose of Eq. (9.5) is taken and is demonstrated to be [4]:

$$[x^{(1)}, \ \dots \ ,x^{(m)}] \begin{bmatrix} H_{11} & \dots & H_{1n} \\ \vdots & \ddots & \vdots \\ H_{m1} & \dots & H_{mn} \end{bmatrix} = \left[y^{(1)}, \ \dots \ ,y^{(m)} \right] \tag{9.6}$$

where m and n are the input and output DOFs, respectively. The operational measurements need to be obtained in unsteady conditions (such as engine run-up or run-down) to obtain a set of synchronized measurement blocks (r) [4]. Using this approach, Eq. (9.6) can be rewritten as [4]:

$$\begin{bmatrix} x_1^{(1)} & \dots & x_1^{(m)} \\ \vdots & \ddots & \vdots \\ x_r^{(1)} & \dots & x_r^{(m)} \end{bmatrix} \begin{bmatrix} H_{11} & \dots & H_{1n} \\ \vdots & \ddots & \vdots \\ H_{m1} & \dots & H_{mn} \end{bmatrix} = \begin{bmatrix} y_1^{(1)} & \dots & y_1^{(n)} \\ \vdots & \ddots & \vdots \\ y_r^{(1)} & \dots & y_r^{(n)} \end{bmatrix} \tag{9.7}$$

Equation (9.6) can be written in a more compact form, defined by:

$$XH = Y \tag{9.8}$$

where the transfer function is determined by:

$$H = X^{-1}Y \tag{9.9}$$

One of the critical weaknesses of OPA is the erroneous data which may be obtained by missing vibration propagation paths in the system. Insufficient time or equipment constrains will limit the number of potentially significant references that will be considered [16].

In the classical TPA mount stiffness approach, missing transfer paths will show up in the analysis as a difference between the target and measured output signal. This has been used as a performance quality indicator of the analysis [16]. However, OPA does not benefit from this phenomenon. The energy propagation effects of the missing paths are carried over to the existing paths, making error detection challenging. Therefore, to perform OPA, a significant degree of attention is required to avoid such issues.

9.2.3 OPAX Method

To overcome limitations faced by OPA, a technique called operation path analysis with exogenous inputs (OPAX) has been developed. This method is as efficient as OPA and accurate as classical TPA. The proposed method utilized a different means in identifying the operational loads. The OPAX technique utilizes parametric models to characterize the operational loads as a direct function of measured inputs to the system, such as the acceleration or velocity inputs to the system [18, 21]. The operational loads are determined using [18]:

1. Measurement of the transfer path input and output signals.
2. The least squares (LS) approach to determine FRFs.

To determine propagation paths from the input to the target location, the following variables need to be measured which are critical to OPAX [18]:

- Measurements:
  - Responses: Acceleration at target location.
  - Path input: Acceleration at the input.
- FRF measurements from the source to the selected target.

In order to perform OPAX, there are five major steps which need to be implemented [18]:

- **Phase 1: Operational measurements**
  This phase is done to obtain the time data of all measurements at the targets at different conditions [18].
- **Phase 2: FRF measurements**
  The FRFs between the input and the output are measured.
- **Phase 3: Identification of parametric load models**
  The loads are determined by a parametric mean, using the data of phases 1 and 2. These parameters are not frequency dependent, decreasing the complexity of the problem. This is one of the main advantages of OPAX. Further information regarding this phase can be obtained by referring to [18].
- **Phase 4: Determination of operational forces**
  All the input loads in the system are determined.
- **Phase 5: Determination of transfer path contributions**
  The path contribution for each receiving location is determined. Using the path contribution results, critical vibration paths can be determined.

9.2.4 Global Transfer Direct Transfer TPA Method

A method that can be implemented to analyze the effects of making altercations to a specific structural pathway by making physical adjustments on the structure under analysis to determine the vibrational characteristics is the Global Transfer Direct Transfer (GTDT) TPA method. This method minimizes the difficulties faced by the original TPA method in

determining path contributions in the system. The main challenge encountered by the original TPA method in identifying path contribution is described as disconnecting the active
component to measure the FRF. The GTDT approach is defined as [22]:

$$x_i = \sum_{j=1, j\neq i}^{N-1} T_{ij}^D x_j + T_{ii}^D x_i^{ext} \tag{9.10}$$

where x_i^{ext} is the response, T_{ij}^D and T_{ii}^D are blocked transmissibilities, and N defines the DOF
in the system. Equation (9.10) shows that the total using the GTDT method is the summa-
tion of DOF responses (x_j) and part of the response that is caused by the force acting directly
on the location.

The difficulties faced by the GTDT method are encountered when determining the
direct transmissibility T_{ij}^D, which means preventing the movement of all DOFs except j
and i when determining T_{ij}^D [22]. One possible way to determine T_{ij}^D would be to physically
attempt to prevent the movement of the determined DOFs in order to determine direct
transmissibility [23]. However, this method suffers from some drawbacks. Another
approach that can be implemented to determine T_{ij}^D is to use standard transmissibility of the system. The key point of this technique is to use standard transmissibility
(global transmissibility) to obtain the desired direct transmissibility. Obtaining the global
transmissibility is easy and does not require any dismounting [22].

9.2.5 Bond Graph TPA Method

Another powerful method based on the transmissibility principles developed by
Mashayekhi and Behdinan [21, 24] can be implemented to perform TPA; it is called
the bond graph TPA method. The theory can be utilized to perform analysis on a range
of engineering systems. The bond graph TPA method uses graphical representation to
determine the dynamic characteristics of the engineering system under analysis, which at
times can be difficult to obtain using other TPA methodologies, as previously explained.
The bond graph TPA method captures the vibration energy propagation between the
components of system [25]. The advanced capabilities of the proposed bond graph TPA
methodology provides a vital tool in performing expensive and labor-intensive design
changes without requiring any manufactured prototypes compared with the previously
outlined TPA methodologies; hence, this significantly reduces development costs [21, 26]
. In addition, the existing TPA methodologies suffer from inaccuracies in the obtained
experimental results due to possible changes in the experimental setup and boundary
conditions in the system, making these methodologies sensitive to slight modifications in
the testing conditions, which can significantly affect the accuracy of the obtained data [21].
The bond graph TPA methodology is a powerful theoretical approach which can be used
to provide researchers a design tool to mitigate and minimize vibration transfer in various
large structures for the lowest possible costs [15]. Due to the theoretical capabilities of
bond graph TPA theory, the methodology has been implemented on a reduced passenger
aircraft aeroengine to study and minimize vibration energy propagation throughout the
structure.

9.3 Bond Graph Formulation

To implement the TPA bond graph methodology, the engineering system under analysis must be decomposed into smaller parts. The smallest part in bond graph methodology is called a component. In bond graph methodology, ports are locations where various subsystems are connected, and power can transfer through [27]. In bond graph methodology, all power variables are categorized into either effort or flow [27]. Table 9.1 illustrates various examples of effort and flow variables.

Power ($P(t)$), expressed as the multiplication of effort and flow variables, is illustrated by the following equation:

$$P(t) = e(t) \cdot f(t) \tag{9.11}$$

In addition, momentum $p(t)$ and displacement $q(t)$ values are important system parameters in bond graph terminology. Momentum is the time integral of the effort:

$$p(t) = \int^t e(t)dt \tag{9.12}$$

Displacement is defined as the time integral of the flow:

$$q(t) = \int^t f(t)dt \tag{9.13}$$

The energy variable in bond graph terminology is the integral of the power, defined in Eq. (9.12). It is defined by the following equation:

$$E(t) = \int^t e(t)dq(t) = \int^t f(t)dp(t) \tag{9.14}$$

9.3.1 Developing Bond Graphs

As previously stated, to perform TPA using bond graph methodology, it is necessary to develop a bond graph representation of the selected engineering system. A bond graph model is composed of components connected by lines through which energy can transfer. This is demonstrated in Figure 9.2.

Each power bond can carry two power variables, namely the effort and flow. Using the model illustrated in Figure 9.2, the flow and effort variables can be calculated using one of

Table 9.1 Power variable (effort and flow).

Effort, $e(t)$	Flow, $f(t)$
Force, $F(t)$ (Newton [N])	Velocity, $V(t)$ (meters per second [m/s])
Torque, $\tau(t)$ (Newton-meter [N-m])	Angular velocity, $\omega(t)$ (radians per second [rad/s])
Pressure, $P_r(t)$	Flow rate, $Q(t)$
Voltage, $v(t)$	Current, $i(t)$

Figure 9.2 Component attachments using power bonds. Source: Adapted from [25].

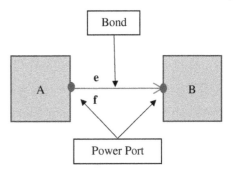

Figure 9.3 Causality (demonstrating the direction of effort and flow). Source: Adapted from [25].

the components in the model [25]. For example, the effort variable can be determined using component A and the flow variable can be determined using component B. Which of the components the effort variable is determined from is illustrated by the use of a stroke called *causality* [25], as shown in Figure 9.3.

There are two forms of junctions which are utilized in the development of bond graphs. The first type is the 0-junction. In the literature, other names used for the 0-junction are flow or common flow junction. An example of a 0-junction is shown for a 3-bond element in Figure 9.4. The 0-junction has the property in which power is not stored or dissipated:

Figure 9.4 Representation of a 3-bond 0-junction element.

$$e_1 f_1 + e_2 f_2 + \ldots + e_n f_n = 0 \tag{9.15}$$

where n defines the number ports of the 0-junction bond.

Moreover, the 0-junction has the property in which all the effort values are equivalent at the node, which is evident by the following equation:

$$e_1(t) = e_2(t) = \ldots = e_n(t) \tag{9.16}$$

Combining Eqs. (9.15, 9.16) yields the following formula:

$$f_1(t) + f_2(t) + \ldots + f_n(t) = 0 \tag{9.17}$$

As shown by Eq. (9.17), using the 0-junction leads to equivalent flow variables but completely different effort variables at the junction

The other category of junction used in the bond graph theory is called the 1-junction. The junction is also referred to as effort or common flow junction. Figure 9.5 represents the symbol for a 1-junction element. As for the 0-junction, the 1-junction is defined in such a way that the power at the node of the element is neither stored nor dissipated. This idea is expressed in Eq. (9.15). Contrary to the 0-junction, the 1-junction is defined by the property in which the flow values at the node of the element are all equivalent,

Figure 9.5 Representation of a 3-bond 1-junction element.

demonstrated by Eq. (9.18):

$$f_1(t) = f_2(t) = \ldots = f_n(t) \tag{9.18}$$

Combining Eqs. (9.15) and (9.18) yields the following formula:

$$e_1(t) + e_2(t) + e_3(t) + \ldots + e_n(t) = 0 \tag{9.19}$$

The sources (inputs) in the bond graph methodology are one of the important aspects in the analysis of a system, represented by S_e and S_f. Input sources are critical to include in any engineering system as they significantly contribute to the dynamic response.

Another important element in developing a bond graph model is the implementation of a 2-bond element, known as transformer (transfer function), which has the symbol TF in bond graphs. The transformers are used to demonstrate the amount of energy flow from one port to another, in applicable scenarios. TFs are used to reflect the fact that only a specified portion of effort or flow is passed from one component to another. The quantity and equation of the TF must be determined based on specific case scenarios.

The following steps demonstrate the overall procedure in developing a mechanical system bond graph [25]:

1. All the velocity components need to be determined and are then represented by a 1-junction.
2. To represent the model, dampers, stiffnesses and TFs are inserted into the bond graph using a 0-junction in between 1-junctions.
3. Any inertia elements in the model are connected to their corresponding 1-junction.
4. Any available source elements are connected to their corresponding 1-junction.
5. Depending on specific case scenarios, a direction for the energy flow must be assigned using half arrows.
6. All the 1-junctions that correspond to a zero velocity must be removed.
7. The bond in bond graph model must be numbered for referencing purposes.
8. The causality for each bond must be assigned.

9.4 Bond Graph Modeling of an Aeroengine

In order to implement the proposed bond graph methodology on an aeroengine model to study vibration energy propagation throughout the system, a reduced aeroengine model is proposed. Aeroengines are composed of multiple cylindrical shaped structures (called casings) and multiple rotor systems along with multiple compressor blades [28]. The casings are connected by bolted flanges. The aeroengine casings are housed in the nacelle.

9.4.1 Reduced Aeroengine Model

The proposed aeroengine model is based on aeroengine schematic illustrated in Figure 9.6 As illustrated in Figure 9.6, the nacelle which houses the casings is attached to the pylon, which is connected to aircraft wing. The pylon plays a critical role in transferring vibration energy from the active vibrating source (rotor system) to the aircraft wing and then fuselage. As previously mentioned, minimizing vibration transfer to the aircraft fuselage is the

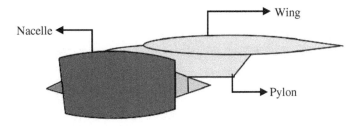

Figure 9.6   Aeroengine schematic.

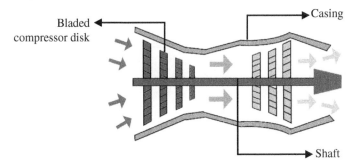

Figure 9.7   Aeroengine cross-section.

primary goal of this study to enhance comfort and safety of the passengers and cabin crew. Figure 9.7 illustrates a 2D cross-section of an aeroengine which consists of the rotor system, casings, and the bladed compressors. The aeroengione casings are assumed to be made of titanium due to high temperature resistance properties [29]. The pylon is also assumed to be made of titanium.

Figure 9.8 represents a proposed lumped representation of an aeroengine model. The aeroengine casing (M_b) represented by a rectangular beam is excited by the cyclic input force of the rotor system due to mass eccentricity caused by limitations and deficiencies in the manufacturing of the rotor system. These forces are transferred to the casings by propagating through two parallel path containing dampers (C_{A1} and C_{A2}) and springs (K_{A1} and K_{A2}) to represent the structural stiffness and damping of the connection. The casing is connected to the aeroengine nacelle through two parallel dampers (C_{P1} and C_{P2}) and springs (K_{P1} and K_{P2}). The nacelle is connected to the wing by pylon attachment points which are represented by two parallel paths of dampers (C_{LW} and C_{LR}) and springs (K_{LW} and K_{RW}), which represent the pylon stiffness and damping characteristics.

9.4.2   Aeroengine Bond Graph

To obtain the governing dynamic equations of motion, the bond graph associated with Figure 9.8 is developed and illustrated in Figure 9.9. After developing the associated bond graph of the reduced aeroengine model, equations of motion associated with the aeroengine structure can be readily obtained. The details for developing bond graphs can be found in

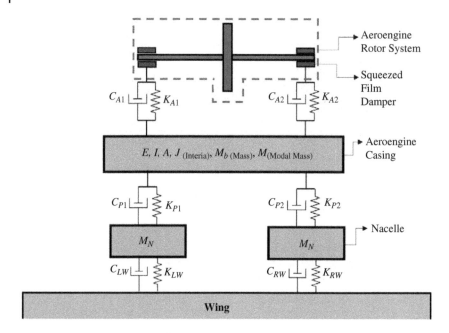

Figure 9.8 Reduced aeroengine mechanical representation.

Section 9.2.2. The utilized aeroengine model parameters and characteristics are presented in Table 9.2.

9.4.3 State Space Equation of the Reduced Aeroengine

The next step after developing bond graphs (which is critical to TPA) is the development of the system equations governing the dynamic system. After obtaining the engineering system bond graph, the state space equation can be readily obtained. This is stated to be one of the main benefits of implementing the bond graph technique.

By analyzing the developed bond graphs, the system is defined in terms of significant variables. The following step-by-step instructions should be followed to determine the state space equation of the system under analysis [27]:

1. The significant input and energy variables in the system must be determined. These selections will show up in the final representative equations. Using the selected and identified significant variables, the first-order set of equations must be formulated.
2. Reformulate the developed first-order equations into compact state space forms

The following equation structure will be determined using the bond graph technique:

$$\dot{x}_1(t) = a_{11}x_1 + a_{12}x_2 + \cdots + a_{1n}x_n + b_{11}u_1 + b_{12}u_2 + \cdots + b_{1r}u_r$$
$$\dot{x}_2(t) = a_{21}x_1 + a_{22}x_2 + \cdots + a_{2n}x_n + b_{21}u_1 + b_{22}u_2 + \cdots + b_{2r}u_r \qquad (9.20)$$
$$\dot{x}_n(t) = a_{n1}x_1 + a_{n2}x_2 + \cdots + a_{nn}x_n + b_{n1}u_1 + b_{n2}u_2 + \cdots + b_{nr}u_r$$

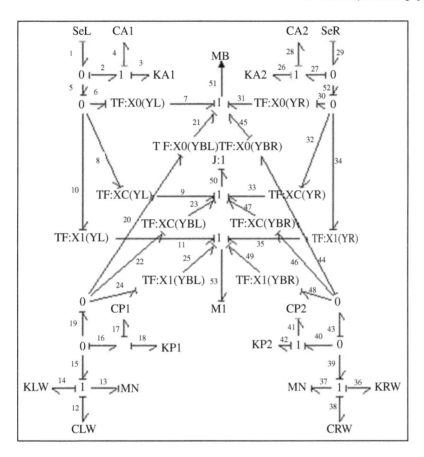

Figure 9.9 Reduced aeroengine bond graph model.

where a_{ij} and b_{ij} are constants. Equation (9.20) can be formulated into a matrix formation for a dynamic system, which is defined as:

$$X = AX + BU$$
$$Y = CX + DU \tag{9.21}$$

where X is the vector of state variables, U is the vector of input, Y is the vector of output, matrix A is composed of coefficients h_{ij} (state matrix), and matrix B is composed of coefficients g_{ij} (input matrix), C is the output matrix, and D is the direct transmission matrix, respectively. Vectors $X, U, Y, A,$ and B are defined by Eq. (9.22):

$$X = [x1\, x2 \cdots xn]^T$$
$$U = [u1\, u2 \cdots ur]^T \tag{9.22}$$
$$Y = [y1\, y2 \cdots yr]^T$$
$$A = \left[\begin{pmatrix} h11 & \cdots & h1n \\ \vdots & \ddots & \vdots \\ hn1 & \cdots & hnn \end{pmatrix} \right]$$

$$B = \left[\begin{pmatrix} g11 & \cdots & g1n \\ \vdots & \ddots & \vdots \\ gn1 & \cdots & gnn \end{pmatrix} \right]$$

Equation (9.21) defines the time domain dynamics equation of motion, yet a frequency domain analysis is more appropriate for a TPA. The mathematical Laplace transform can be implemented to transfer Eq. (9.21) into the frequency domain, as shown:

$$X = (sI - A)^{-1}BU$$
$$Y = [C(sI - A)^{-1}B + D]U \tag{9.23}$$

9.4.4 Sample Calculation: Output and Direct Transmissibility Matrices

The state-space equation of the study case can be obtained using the methodology outlined in Section 9.4. Sample calculations are provided to illustrate the mathematical procedure to obtain the state variables $\dot{q}3$, $\dot{q}14$; corresponding to the aeroengine bond graph presented in Figure 9.9. Once A and B are determined, they are consistent and do not change based on

Table 9.2 Reduced aeroengine model parameters.

Structure	Characteristic	Assigned parameters
Aeroengine casing	M_b	225 kg
	Density (ρ)	4500 kg/m^3
	Effective modal mass $(M_1$, modal)	182 kg
	Mass moment of inertia $(M_1$, inertia)	18.8 kg/m^2
	Length	1 m
	Width	1 m
	Height	0.05 m
Aeroengine nacelle	M_N	200 kg
The structural connections in the aeroengine	K_{A1}	3×10^4 N/m
	K_{A2}	3×10^4 N/m
	K_{P1}	5×10^6 N/m
	K_{P2}	5×10^6 N/m
	K_{LW}	3×10^5 N/m
	K_{RW}	3×10^5 N/m
	C_{A1}	4 N.s/m
	C_{A2}	4 N.s/m
	C_{P1}	3.5 N.s/m
	C_{P2}	3.5 N.s/m
	C_{LW}	3 N.s/m
	C_{RW}	3 N.s/m

different case studies, yet C and D should be formulated based on specific case scenarios. Initially, to calculate the state variable $\dot{q}3$, the flow into bonds 2 and 3 are defined as:

$$\dot{q}3 = \dot{q}4 = \dot{q}2 = f2 = f3 = \frac{e4}{C_{A1}} \tag{9.24}$$

Effort variable for bond 3 can be defined as:

$$e3 = K_{A1} \cdot q3 \tag{9.25}$$

Effort into bond 2 can be defined by:

$$e2 = e4 + e3 = Sel \tag{9.26}$$

Mathematical reformulation would lead to:

$$e4 = Sel - \frac{e3}{K_{A1}} \tag{9.27}$$

Combining Eqs. (9.24) and (9.25), and Eq. (9.27) would yield the flow at bond 3 of the aeroengine:

$$\dot{q}3 = \frac{1}{C_{A1}}(Sel - q3 \cdot K_{A1}) \tag{9.28}$$

The following equation yields the term defining the state variable $\dot{q}14$:

$$\dot{q}14 = \dot{q}12 = \dot{q}13 = \dot{q}15 = f14 = f12 = f13 = f15 = \frac{P13}{M_N} \tag{9.29}$$

The state variables vector of the aeroengine model (x) is defined by Eq. (9.30). The remaining state variables are defined by Eq. (9.31).

$$x = [\dot{q}3 \ \dot{q}14 \ \dot{P}13 \ \dot{q}18 \ \dot{q}26 \ \dot{q}36 \ \dot{P}37 \ \dot{q}42 \ \dot{P}50 \ \dot{P}51 \ \dot{P}52 \ \dot{q}53]^T \tag{9.30}$$

$$\begin{bmatrix} Q \\ P \end{bmatrix} = \begin{bmatrix} A_{11} & A_{12} \\ A_{21} & A_{22} \end{bmatrix} \cdot \begin{bmatrix} \overline{Q} \\ \overline{P} \end{bmatrix} + \begin{bmatrix} B1 \\ B2 \end{bmatrix} \tag{9.31}$$

Matrices Q and P are defined as:

$$[Q] = [\dot{q}_3 \ \dot{q}_{14} \ \dot{q}_{18} \ \dot{q}_{26} \ \dot{q}_{36} \ \dot{q}_{42} \ \dot{q}_{53}]^T \tag{9.32}$$

$$[P] = [\dot{P}13 \ \dot{P}37 \ \dot{P}50 \ \dot{P}51 \ \dot{P}52]^T \tag{9.33}$$

Matrices $\overline{Q}$ and $\overline{P}$ are defined as:

$$[\overline{Q}] = [q_3 \ q_{14} \ q_{18} \ q_{26} \ q_{36} \ q_{42} \ q_{53}]^T \tag{9.34}$$

$$[\overline{P}] = [P13 \ P37 \ P50 \ P51 \ P52]^T \tag{9.35}$$

Matrices $\boldsymbol{A}_{11}$ and $\boldsymbol{A}_{12}$ are defined as:

$$\boldsymbol{A}_{11} = \begin{bmatrix} \dfrac{-K_{A1}}{C_{A1}} & 0 & 0 & 0 & 0 & 0 & 0 \\[2mm] 0 & 0 & \dfrac{1}{M_N} & 0 & 0 & 0 & 0 \\[2mm] 0 & 0 & \dfrac{-1}{M_N} & 0 & 0 & 0 & 0 \\[2mm] 0 & 0 & 0 & 0 & \dfrac{-K_{A2}}{C_{A2}} & 0 & 0 \\[2mm] 0 & 0 & 0 & 0 & 0 & 0 & \dfrac{1}{M_N} \\[2mm] 0 & 0 & 0 & 0 & 0 & 0 & \dfrac{-1}{M_N} \\[2mm] 0 & 0 & 0 & 0 & 0 & 0 & 0 \end{bmatrix}$$

$$\boldsymbol{A}_{12} = \begin{bmatrix} 0 & 0 & 0 & 0 & 0 \\[2mm] 0 & 0 & 0 & 0 & 0 \\[2mm] 0 & \dfrac{-1}{X0(YBL)M_B} & \dfrac{-1}{XC(YBL)J} & \dfrac{-1}{X1(YBL)M_1} & 0 \\[2mm] 0 & 0 & 0 & 0 & 0 \\[2mm] 0 & 0 & 0 & 0 & 0 \\[2mm] 0 & \dfrac{-1}{X0(YBR)M_B} & \dfrac{-1}{XC(YBR)J} & \dfrac{-1}{X1(YBR)M_1} & 0 \\[2mm] 0 & 0 & 0 & \dfrac{1}{M1} & 0 \end{bmatrix} \tag{9.36}$$

Matrices $\boldsymbol{A}_{21}$ and $\boldsymbol{A}_{22}$ are defined as:

$$\boldsymbol{A}_{21} = \begin{bmatrix} 0 & -K_{LW} & \dfrac{-(C_{P1}+C_{LW})}{M_N} & K_{P1} & 0 & 0 & 0 \\[2mm] 0 & 0 & 0 & 0 & 0 & -K_{RW} & \dfrac{-(C_{P2}+C_{RW})}{M_N} \\[2mm] 0 & 0 & \dfrac{C_{P1}}{XC(YBL)MN} & \dfrac{K_{P1}}{XC(YBL)} & 0 & 0 & \dfrac{-C_{P2}}{XC(YBR)M_N} \\[2mm] 0 & 0 & \dfrac{C_{P1}}{X0(YBL)M_N} & \dfrac{K_{P1}}{X0(YBL)} & 0 & 0 & \dfrac{-C_{P2}}{X0(YBR)M_N} \\[2mm] 0 & 0 & \dfrac{-C_{P1}}{X1(YBL)M_N} & \dfrac{K_{P1}}{X1(YBL)} & 0 & 0 & \dfrac{-C_{P2}}{X1(YBR)M_N} \end{bmatrix}$$

$$A_{22} = \begin{bmatrix} 0 & \dfrac{C_{P1}}{X0(YBL)M_B} & \dfrac{-C_{P1}}{XC(YBK)J} & \dfrac{-C_{P1}}{X1(YBL)M_1} & 0 \\[2ex] -K_{P2} & \dfrac{-C_{P2}}{X0(YBR)M_B} & \dfrac{-C_{P2}}{XC(YBR)J} & \dfrac{-C_{P2}}{X1(YBR)M_1} & 0 \\[2ex] \dfrac{K_{P2}}{XC(YBR)} & \dfrac{-C_{P2}}{X0(YBR)M_B} & \dfrac{-C_{P2}}{XC(YBR)J} & \dfrac{-C_{P2}}{X1(YBR)M_1} & 0 \\[2ex] & \dfrac{-C_{P1}}{X0(YBL)M_B} & \dfrac{-C_{P1}}{XC(YBL)J} & \dfrac{C_{P1}}{X1(YBL)M_1} & \\[2ex] \dfrac{K_{P2}}{X0(YBR)} & \dfrac{-C_{P2}}{X0(YBR)M_B} & \dfrac{-C_{P2}}{XC(YBR)J} & \dfrac{-C_{P2}}{X1(YBR)M_1} & 0 \\[2ex] & \dfrac{C_{P1}}{X0(YBL)M_B} & \dfrac{C_{P1}}{XC(YBL)J} & \dfrac{C_{P1}}{X1(YBL)M_1} & \\[2ex] \dfrac{K_{P2}}{X1(YBR)} & \dfrac{-C_{P2}}{X0(YBR)M_B} & \dfrac{-C_{P2}}{XC(YBR)J} & \dfrac{-C_{P2}}{X1(YBR)M_1} & 0 \\[2ex] & \dfrac{CP1}{X0(YBL)M_B} & \dfrac{C_{P1}}{XC(YBL)J} & \dfrac{C_{P1}}{X1(YBL)M_1} & \end{bmatrix} \quad (9.37)$$

Matrix **B1** is defined as:

$$[B1] = \begin{bmatrix} 0 \\ 0 \\ 0 \\ \dfrac{Ser}{C_{A2}} \\ 0 \\ 0 \\ 0 \end{bmatrix} \quad (9.38)$$

Matrix **B2** is defined as:

$$[B2] = \begin{bmatrix} 0 \\[1ex] 0 \\[1ex] \dfrac{Sel}{X0(YR)} + \dfrac{Ser}{X0(YL)} \\[2ex] \dfrac{Sel}{XC(YL)} + \dfrac{Ser}{XC(YR)} \\[2ex] \dfrac{Sel}{X1(YL)} + \dfrac{Ser}{X1(YR)} \end{bmatrix} \quad (9.39)$$

9.5 Transmissibility Principle

After determining the governing equation of motions, the transmissibility between various aeroengine elements (inertia elements) can be determined. This is achieved by implementing the theory of global transmissibility by definition, the global transmissibility between two elements (i.e. inertia mass elements) mi and mj is defined as the relative displacements between them when an external load is exerted on mi when no other restrictions are imposed on the overall system [24].

$$TR_{ij}^G = \frac{X_j}{X_i}\bigg|_{|Fi|=1} \tag{9.40}$$

Equation (9.40) is reformulated to the frequency domain by [24]:

$$TR_{ij}^G = \frac{SX_j}{SX_i} = \frac{f^{(j)}}{f^{(i)}} \tag{9.41}$$

where $f^{(i)}$ demonstrates the flow in the respective element. $f^{(i)}$ can be defined by [24]:

$$f^{(i)} = [C^i(sI - A)^{-1}B + D^i]U \tag{9.42}$$

This would lead to:

$$TR_{ij}^G = \frac{[C^j(sI - A)^{-1}B + D^j]U}{[C^i(sI - A)^{-1}B + D^i]U} \tag{9.43}$$

Using Eq. (9.43), the global transmissibility can be determined using the Bode magnitude function which shows the frequency response of the transmissibility at various frequencies [24]:

$$TR_{ij}^G = \frac{Bodemag(A, b, C^j, D^j)U}{Bodemag(A, b, C^i, D^i)U} \tag{9.44}$$

9.6 Bond Graph Transfer Function

The TF $X_0(y)$ and $X_1(y)$ depicted in the reduced aeroengine bond graph represents the modal shape deflection for the first and second natural frequencies at various locations on the casing. The free–free transverse vibration of the casing can be represented by:

$$\frac{d^4Y}{dX^4} - \frac{\rho A}{EI}\omega^2 Y = 0 \tag{9.45}$$

where ρ is the density, A is the cross-sectional area, E is Young's modulus of elasticity, and I is the second moment of area. By substituting $\varphi^4 = \frac{\rho A}{EI}\omega^2$, Eq. (9.45) can be reduced to:

$$\frac{d^4Y}{dX^4} - \varphi^4 Y = 0 \tag{9.46}$$

The separation of variables is as follows:

$$Y(x, t) = w(x)u(t) \tag{9.47}$$

Using the separation of variables, the solution for the spatial part of Eq. (9.35) can be defined as:

$$w(x) = A1\sin(\varphi x) + A2\cos(\varphi x) + A3\sinh(\varphi x) + A4\cosh(\varphi x) \qquad (9.48)$$

where A is determined based on the initial conditions of the system.

The eigenfunction of the casing corresponding to the bond graph transfer function is determined and defined as:

$$\omega_n(x) = A_n\left[\sin\phi_n x - \sinh\phi_n x - \frac{\sin\phi_n L + \sinh\phi_n L}{\cos\phi_n L + \cosh\phi_n L}(\cos\phi_n x - \cosh\phi_n x)\right] \qquad (9.49)$$

9.7 Aeroengine Global Transmissibility Formulation

By utilizing Eq. (9.44), the transmissibility between various inertia elements in the reduced aeroengine model can be determined. For design and analysis purposes, the global transmissibility between active vibrating component in the aeroengine (*rotor system*, bond 27) and the aeroengine nacelle (M_N, bond 37/39) is determined. The state matrix $\mathbf{A}$ and input matrix $\mathbf{B}$ were previously determined and are illustrated by Eq. (9.31). The following calculations illustrate the procedure in obtaining matrix $\mathbf{C}$ and $\mathbf{D}$. The global transmissibility is defined as $T^G_{39,27} = \frac{f39/Se_R}{f27/Se_R}$.

$$f^{(i)} = [C^i(sI - A)^{-1}B + D^i]U \qquad (9.50)$$

$$\frac{\vec{Y}}{\vec{U}} = [C^{39}(sI - A)^{-1}B + D^{39}] \qquad (9.51)$$

$$Y39 = f39 = \frac{P37}{MN} = CX + D\vec{U} \qquad (9.52)$$

$$Y39 = C^{39}X + D^{39}\vec{U} = \left[0\,0\,0\,0\,0\,0\frac{1}{MN}0\,0\,0\,0\,0\right][0\,0\,0\,0\,0\,0\,P_{37}\,0\,0\,0\,0\,0]^{\mathrm{T}} + [0]Se r\vec{U} \qquad (9.53)$$

The output matrix $\mathbf{C}$ and direct transmission matrix $\mathbf{D}$ for bond 27 are determined as:

$$f^{(i)} = [C^i(sI - A)^{-1}B + D^i]U \qquad (9.54)$$

$$\frac{\vec{Y}}{\vec{U}} = [C^{27}(sI - A)^{-1}B + D^{27}] \qquad (9.55)$$

$$Y27 = f27 = \frac{e28}{CA2} = \frac{Ser - q29 \cdot KA2}{CA2} = CX + D\vec{U} : \qquad (9.56)$$

$$Y27 = C^{27}X + D^{27}\vec{U} = \left[0\,0\,0\,0\,\frac{-KA2}{CA2}\,0\,0\,0\,0\,0\,0\,0\right][0\,0\,0\,0\,q_{29}\,0\,0\,0\,0\,0\,0\,0]^{\mathrm{T}}$$
$$+ \left[\frac{1}{CA2}\right]Ser\vec{U} \qquad (9.57)$$

Using Eq. (9.44), global transmissibility can be defined as:

$$T^G_{39,27} = \frac{\mathrm{Bodemag}(\mathbf{A}, \mathbf{B}, \mathbf{C}_{39}, \mathbf{D}_{39})}{\mathrm{Bodemag}(\mathbf{A}, \mathbf{B}, \mathbf{C}_{27}, \mathbf{D}_{27})} \qquad (9.58)$$

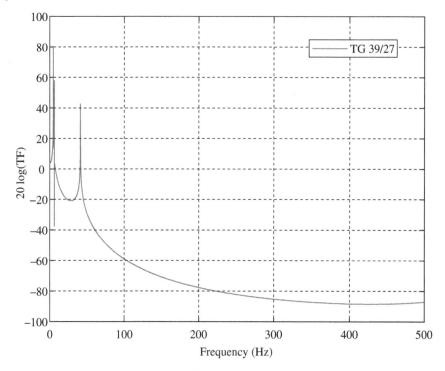

Figure 9.10 Global transmissibility $T^{G.}_{39,27}$.

The global transmissibility $T^{G}_{39,27}$ for the frequency range of 0–500 Hz is demonstrated by Figure 9.10. Figure 9.10 demonstrates the global transmissibility between the rotor system which is the active vibrating component in the aeroengine and the receiving passive structure, defined to be the nacelle. The global transmissibility graph illustrates three transmissibility peaks that are of interest. The main transmissibility peaks are 4.8, 6, and 41.5 Hz in the frequency range of analysis (0–500 Hz). As is evident, using bond graph methodology allows the rapid determination of dominant frequency peaks in the frequency range of interest without any experimental measurements.

In the second case study, the transmissibility between the casing (*Mb*, bond 51) and the nacelle (*MN*, bond 13) is determined. The following calculations illustrate the procedure in obtaining matrices **C** and **D**. The global transmissibility is defined as $T^{G}_{51,13} = \frac{f51/Se_R}{f13/Se_R}$.

$$f^{(i)} = [C^i(sI - A)^{-1}B + D^i]U \tag{9.59}$$

$$\frac{\vec{Y}}{\vec{U}} = [C^{51}(sI - A)^{-1}B + D^{51}] \tag{9.60}$$

$$Y51 = f51 = \frac{P51}{MB} = CX + D\vec{U} : \tag{9.61}$$

$$Y51 = C^{51}X + D^{51}\vec{U} = \left[0\,0\,0\,0\,0\,0\,0\,0\,\frac{1}{MB}\,0\,0\,0\right][0\,0\,0\,0\,0\,0\,0\,0\,P_{51}\,0\,0\,0]^T + [0]Ser\vec{U} \tag{9.62}$$

Matrices **C** and **D** for bond 13 are determined as:

$$f^{(i)} = [C^i(sI - A)^{-1}B + D^i]U \qquad (9.63)$$

$$\frac{\vec{Y}}{\vec{U}} = [C^{13}(sI - A)^{-1}B + D^{13}] \qquad (9.64)$$

$$Y13 = f13 = \frac{P13}{MN} = CX + D\vec{U} : \qquad (9.65)$$

$$Y27 = C^{13}X + D^{13}\vec{U} = \left[0\,0\,\frac{1}{MN}\,0\,0\,0\,0\,0\,0\,0\,0\right][0\,0\,P_{13}\,0\,0\,0\,0\,0\,0\,0\,0]^T + [0]Ser\vec{U} \qquad (9.66)$$

Using Eq. (9.44), global transmissibility is defined as:

$$T^G_{51,13} = \frac{\text{Bodemag}(A, B, C_{51}, D_{51})}{\text{Bodemag}(A, B, C_{13}, D_{13})} \qquad (9.67)$$

The global transmissibility $T^G_{51,13}$ for the frequency range of 0–500 Hz is demonstrated by Figure 9.11. Figure 9.11 demonstrates the global transmissibility between the rotor system (active vibrating component in the aeroengine) and the receiving passive structure (the nacelle). The global transmissibility graph illustrates three transmissibility peaks that are of interest. The main transmissibility peaks are 6, 76 and 427 Hz in the frequency range of analysis (0–500 Hz). The obtained frequency peaks using bond graph methodology are of great importance in the design of aeroengines. It is paramount to ensure that the obtained peaks do not coincide with the operational frequency of the aeroengine as this could cause amplification of vibration transfer in the aeroengine. Therefore, it is required to attenuate

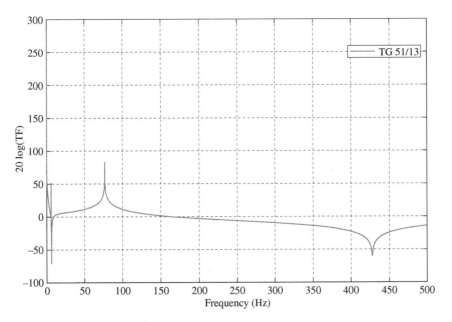

Figure 9.11 Global transmissibility $T^G_{51,13}$.

the identified transmissibility peaks to avoid such amplifications by proposing structural design modifications in the aeroengine. The obtained transmissibility graph illustrated the strength of the bond graph technique in analyzing vibration transfer that becomes evident compared with other TPA techniques as no experimentation or prototypes are required, minimizing development costs in the industry.

9.8 Design Guidelines to Minimize Vibration Transfer

As previously stated, the bond graph method is a powerful theoretical method which can be implemented prior to any aeroengine prototyping. The bond graph method can be implemented before any manufacturing to study and minimize vibration transfer in the aeroengine. It is a design goal to minimize the overall vibration transfer from the aeroengine to the aircraft, such as the transfer of that vibration to the fuselage. The design goal is to minimize vibration transfer by attenuating the transmissibility magnitude peaks, as previously discussed. This can be achieved by proposing structural design modifications on the attachment points that connect various aeroengine components, such as the stiffness and damping parameters. It is assumed that the aeroengine components themselves are a design constraint and can be varied.

For the initial design analysis, the structural damping and stiffness connecting the rotor to the casing will be analyzed $(K_{A1}, K_{A2}, C_{A1}, C_{A2})$ on the global transmissibility between the rotor system and the nacelle. The stiffness parameters will be increased by a factor of 20 and 50% to analyze the overall effects on global transmissibility $(T^G_{39,27})$. The new stiffness parameters used for this study are $K_{A1} = K_{A2} = 3.6 \times 10^4$ N/m and $K_{A1} = K_{A2} = 4.5 \times 10^4$ N/m. In addition, new damping parameters are also used to study the effects on the vibration transfer. The new damping parameters used in the model are $C_{A1} = C_{A2} = 40$ N.s/m and $C_{A1} = C_{A2} = 0.1$ N.s/m. Figures 9.12–9.15 illustrate the effects of the stiffness and damping parameters on $T^G_{39,27}$.

As evident from Figures 9.12–9.15, variation in the damping and stiffness parameters for the attachment points between the rotor system and the nacelle does not illustrate a critical role in attenuating transmissibility peak or shift in frequency peaks. Therefore, as a design guideline, it can be concluded that changing the parameters for the structural attachment between the casing and the rotor system does count as a critical design parameter.

For the second design analysis, the structural damping and stiffness that connects the casing to the nacelle will be analyzed $(K_{P1}, K_{P2}, C_{P1}, C_{P2})$ on the global transmissibility between the rotor system and the nacelle. The stiffness parameters will be increased by a factor of 20 and 50% to analyze the overall effects on global transmissibility $(T^G_{39,27})$. The new stiffness parameters used for this study are $K_{P1} = K_{P2} = 6 \times 10^6$ N/m and $K_{P1} = K_{P2} = 7.5 \times 10^6$ N/m. In addition, new damping parameters are also used to study the effects on the vibration transfer. The new damping parameters used in the model are $C_{P1} = C_{P2} = 35$ N.s/m and $C_{P1} = C_{P2} = 0.1$ N.s/m. Figures 9.16 and 9.17 represent the effects of the stiffness and damping parameters on $T^G_{39,27}$.

As it can be observed from Figure 9.16, increasing K_{P1} and K_{P2} stiffness parameters shows a decreasing trend in transmissibility magnitude peaks and an overall shift in frequency (approximately 5% reduction in peak magnitude for the second transmissibility peak). This

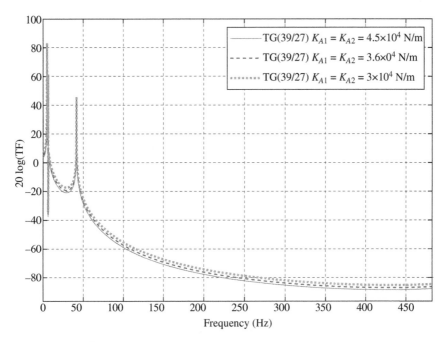

Figure 9.12 $T^G_{39,27}$ for changes in stiffness parameters (K_{A1} and K_{A2}).

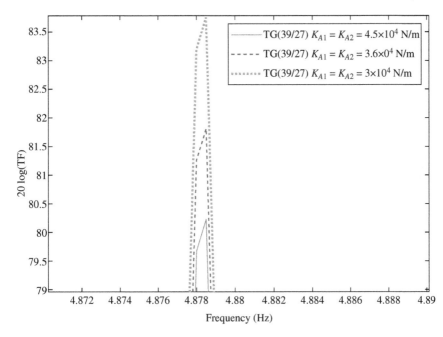

Figure 9.13 First transmissibility peak-$T^G_{39,27}$ for changes in stiffness parameters (K_{A1} and K_{A2}).

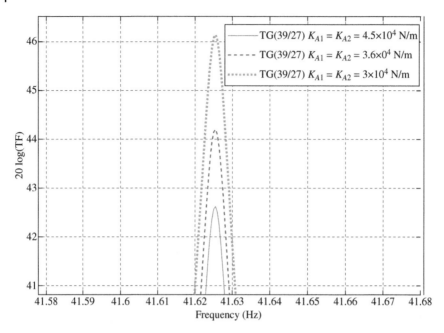

Figure 9.14 Second transmissibility peak-$T^G_{39,27}$ for changes in stiffness parameters (K_{A1} and K_{A2}).

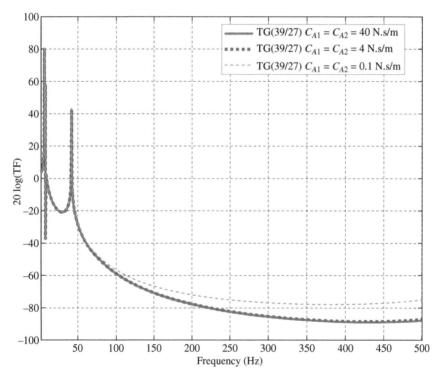

Figure 9.15 $T^G_{39,27}$ for changes in damping parameters (C_{A1} and C_{A2}).

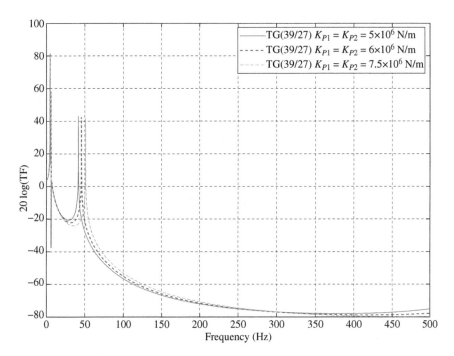

Figure 9.16 $T^G_{39,27}$ for changes in stiffness parameters (K_{P1} and K_{P2}).

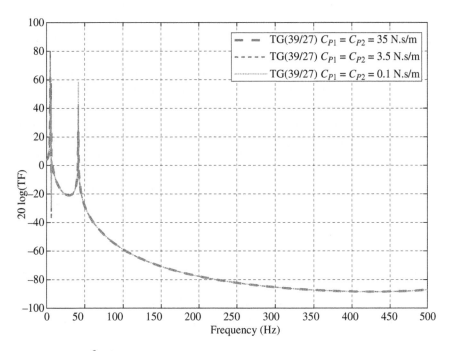

Figure 9.17 $T^G_{39,27}$ for changes in damping parameters (C_{P1} and C_{P2}).

suggests that by increasing the stiffness parameter regarding the casing and the nacelle connection, vibration transfer to the fuselage can be attenuated. Therefore, it is proposed as an aeroengine design guideline to increase the stiffness values of the connection points between the casing and the nacelle to attenuate vibration transfer in the aeroengine structure. In addition, the stiffness parameter can be used to shift the critical transmissibility frequency to the lower or higher ranges as needed to avoid overlapping with critical operational frequency of the aeroengine, which could lead to catastrophic results.

As it can be observed from Figure 9.17, increasing C_{P1} and C_{P2} damping parameters illustrates a significant drop in transmissibility magnitude peaks (approximately 60% reduction in peak magnitude for the second transmissibility peak). Therefore, as a design guideline, the damping parameter of the attachment point between the casing and the nacelle can play a critical role in attenuating vibration transfer in the aircraft. To achieve the proposed design strategies proposed by this analysis, which would require changes in stiffness and damping values, special attention should be placed on selecting appropriate materials that satisfy these needs. For instance, Young's modulus is an important design parameter in material selection, as it has a direct impact on the stiffness of the connection points.

9.9 Conclusion

In this research, the bond graph TPA method is used to perform vibration propagation analysis on the proposed reduced aeroengine model. For vibration TPA, the aeroengine rotor system was determined to be the active vibrating source in the system caused by mass eccentricity. The vibration energy is propagated through the aeroengine casings and then to the pylon and the wing. Using the transmissibility principles along with the bond graph TPA methodology, the global transmissibility between various inertia elements in the system were determined over a selected frequency range. Using this method, aeroengine design guidelines which would minimize the overall vibration transfer in the aeroengine structure were determined. By proposing strategies in minimizing the vibration transfer, it was determined that the mechanical attachments and connections between the casings, nacelle, and the wing play a critical role in attenuating vibration. It has been determined that based on design requirements, it is critically important to accurately select an appropriate Young's modulus value that would result in the specified stiffness values. Implementation of the bond graph TPA method has been demonstrated to be the perfect tool for use in the aerospace industry as it provides a fast and reliable theoretical mean to perform vibration TPA at the lowest cost, which can significantly reduce the operating costs in the sector.

References

1 Liu, Y., Zhang, M., Sun, C. et al. (2019). A method to minimize stage-by-stage initial unbalance in the aero engine assembly of multistage rotors. *Aerosp. Sci. Technol.* 85: 270–276.
2 Sakhaei, B. and Durali, M. (2014). Vibration transfer path analysis and path ranking for NVH optimization of a vehicle interior. *Shock Vib.* https://doi.org/10.1155/2014/697450.

3 Lee, S.-K., Lee, S.-Y., and Sung, S.H. (2016). *Engineering Vibroacoustic Analysis: Methods and Applications*, 1e. Wiley.

4 De Klerk, D. and Ossipov, A. (2010). Operational transfer path analysis: theory, guidelines and tire noise application. *Mech. Syst. Signal Process* 24: 1950–1962.

5 Elliott, A.S., Moorhouse, A.T., Huntley, T., and Tate, S. (2013). In-situ source path contribution analysis of structure borne road noise. *J. Sound Vib.* 332: 6276–6295.

6 De Klerk, D. and Rixen, D.J. (2010). Component transfer path analysis method with compensation for test bench dynamics. *Mech. Syst. Signal Process* 24: 1693–1710.

7 de Klerk, D. (2009). Application of operational transfer path analysis on a classic car. *Daga* 2009: 4–7.

8 Molina-Viedma, Á.J., López-Alba, E., Felipe-Sesé, L. et al. (2018). Modal parameters evaluation in a full-scale aircraft demonstrator under different environmental conditions using HS 3D-DIC. *Materials (Basel)* https://doi.org/10.3390/ma11020230.

9 Viscardi, M. and Arena, M. (2019). Sound proofing and thermal properties of an innovative viscoelastic treatment for the turboprop aircraft fuselage. *CEAS Aeronaut. J.* 10: 443–452.

10 Wright, R.I. and Kidner, M.R.F. (2004). Vibration absorbers: a review of applications in interior noise control of propeller aircraft. *J. Vib. Control* 10: 1221–1237.

11 De Fenza, A., Arena, M., and Lecce, L. (2018). Innovative passive multifrequency propeller device for noise and vibration reduction in turboprop fuselage. *MATEC Web Conf.* 233: 00030. https://doi.org/10.1051/matecconf/201823300030.

12 Van Der Seijs, M.V., De Klerk, D., and Rixen, D.J. (2016). General framework for transfer path analysis: history, theory and classification of techniques. *Mech. Syst. Signal Process* 68–69: 217–244.

13 Gajdatsy, P., Janssens, K., Gielen, L. et al. (2008). Critical assessment of operational path analysis: mathematical problems of transmissibility estimation. *J. Acoust Soc. Am.* 123: 3869.

14 Gloth, G. and Sinapius, M. (2004). Analysis of swept-sine runs during modal identification. *Mech. Syst. Signal Process* 18: 1421–1441.

15 Goege, D. (2007). Fast identification and characterization of nonlinearities in experimental modal analysis of large aircraft. *J. Aircr.* 44: 399–409.

16 Janssens, K., Gajdatsy, P., and Van der Auweraer, H. (2008). Operational Path Analysis: a critical review. ISMA Conference, Leuven.

17 Van de Ponseele, P., Van der, Auweraer, H., and Janssens, K. Source-Transfer-Receiver approaches: a review of methods. In: *Proceedings of ISMA2012-USD2012*, 3645–3658. KU Leuven.

18 Janssens, K., Gajdatsy, P., Gielen, L. et al. (2011). OPAX: a new transfer path analysis method based on parametric load models. *Mech. Syst. Signal Process* 25: 1321–1338.

19 Ewins, D. and Liu, W. (1998). Transmissibility properties of MDOF systems. 16th IMAC Conference, Santa Barbara, CA.

20 Maia, N.M.M., Silva, J.M.M., and Ribeiro, A.M.R. The transmissibility concept in multi-degree-of- freedom systems. *Mech. Syst. Signal Process* 15: 129–137.

21 Mir-Haidari, S.E. and Behdinan, K. (2019). On the vibration transfer path analysis of aero-engines using bond graph theory. *Aerosp. Sci. Technol.* 95: 105516.

22 Guasch, O., García, C., Jové, J., and Artís, P. (2013). Experimental validation of the direct transmissibility approach to classical transfer path analysis on a mechanical setup. *Mech. Syst. Signal Process* 37: 353–369.

23 Mixson, J. and Wilby, J. (1995). Aeroacoustics of flight vehicles. Theory and practice. Noise control. *J. Acoust. Soc. Am.* II: 414–420.

24 Mashayekhi, M.J. and Behdinan, K. (2015). Analytical transmissibility based transfer path analysis for multi-energy-domain systems using bond graphs. *J. Vib. Control* https://doi.org/10.1177/1077546317696362.

25 Borutzky, W. (2006). Bond graph modelling and simulation of mechatronic systems. An introduction into the methodology. 20th European Conference on Modelling and Simulation, Bonn.

26 Mir-Haidari, S.-E. and Behdinan, K. (2019). *Aero-Engine Vibration Propagation Analysis Using Bond Graph Transfer Path Analysis and Transmissibility Theory*. ASME. https://doi.org/10.1115/IMECE2019-10773.

27 Karnopp, D.C., Margolis, D.L., and Rosenberg, R.C. (2006). *System Dynamics*, 4e. Wiley.

28 Böswald, M., Link, M., and Meyer, S. (2003). Experimental and analytical investigations of non-linear cylindrical casing joints using base excitation testing. In: *Conference & Exposition on Structural Dynamics*, 9. Society for Experimental Mechanics.

29 Bettebghor, D., Blondeau, C., Toal, D., and Eres, H. (2013). Bi-objective optimization of pylon-engine-nacelle assembly: weight vs. tip clearance criterion. *Struct. Multidiscip. Optim.* 48: 637–652.

10

Structural Health Monitoring of Aeroengines Using Transmissibility and Bond Graph Methodology

Seyed Ehsan Mir-Haidari and Kamran Behdinan

Department of Mechanical & Industrial Engineering, Advanced Research Laboratory for Multifunctional Lightweight Structures (ARL-MLS), University of Toronto, Toronto, Canada

10.1 Introduction

In recent years, due to increase in demand for structural safety in various engineering disciplines such as aerospace, researchers have focused on developing vibration-based methodologies for monitoring and detecting structural damage in various structures. Safety and performance of an aging structure is a prominent issue, occurring in aircraft, engines, and public infrastructures. There have been significant recent developments in the field of structural health monitoring (SHM) [1–9]. SHM provides a means to monitor structural safety in a continuous manner. The main objective of SHM can be defined as the assessment and monitoring of the structural safety and operational condition to detect any anticipated damage or faults that are beginning to occur or ones that have already propagated and advanced in the structure to prolong the operational lifetime of the structure in a safe manner. This goal is achieved while reducing detection and maintenance costs to significantly enhance and improve on the safety of the structure by a providing an early detection tool to detect structural damage. Moreover, aside from the improved safety and performance, the obtained knowledge from SHM can be utilized to improve on the early designs of the structure. Structural aging is caused by continuous loading conditions (low and high loading forces), leading to fatigue cracks in the structure [10]. The exposure to continuous loading conditions leads to crack growth in the structure, leading to catastrophic failures. This is shown by Figure 10.1.

As evident by Figure 10.1, crack follows an exponential growth [10]. At the first stages, only a small crack is detected with size S_0. However, as the cyclic load continues, the crack growth follows an exponential order, making the growth unstable (S_{cri}) which leads to catastrophic failure in the structure.

SHM can be categorized into five main phases that need to be addressed to perform a comprehensive structural health assessment [3, 4, 10–15]. This process is outlined in Figure 10.2.

It must be emphasized that the most complicated phase in performing SHM is the prediction phase, where the remaining operational life of the structure is determined.

Advanced Multifunctional Lightweight Aerostructures: Design, Development, and Implementation,
First Edition. Kamran Behdinan and Rasool Moradi-Dastjerdi.

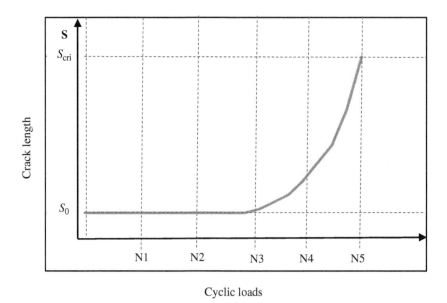

Figure 10.1 Crack growth under continuous loading conditions. Source: Adapted from [10].

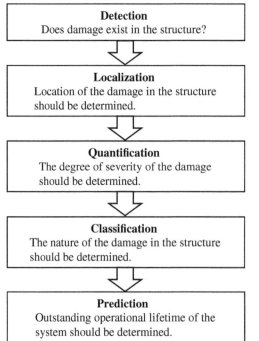

Figure 10.2 Description of assessment phases in structural health monitoring.

To perform this step, the global SHM results need to be combined with damage models such as fracture mechanics models to evaluate the crack propagation in the structure to determine the remaining useful lifetime of the structure. Research articles regarding lifetime assessment based on damage analysis and crack growth analysis can be found in the literature [16–19].

In general, there are two SHM methods that can be implemented. These methods are called local and global methods. Local methods are implemented on small and localized areas of the structure. Using this method, nondestructive experimental methods such as ultrasound, X-rays, and gamma rays are used. The local method requires the accessibility of the area being inspected, which may not be possible in all cases and applications. The local method allows for quick localization of the damage and provides high sensitivity which can be used to determine small amounts of damage in the structure. The global SHM method relies on the theory that the overall global behavior of the structure relies on the localized damage and defects [20–22]. These structural defects lead to changes in the vibrational characteristics of the system, which can be used to detect the presence of defects and deficiencies in the structure.

For an engineering structure, damage in the system leads to changes in the vibrational characteristics and dynamic behavior of the system, such as changes in the mode shapes, eigenfrequencies, damping and stiffness values of the structure [23]. An accurate protocol in measuring variations and changes of the dynamic characteristics leads to an accurate measurement of the damage in the structure. The vibrational characteristics can be obtained by performing structural dynamic analysis on the structure. The variations and changes of the modal characteristics due to damage is not consistent for all modes [3, 24]. The specific changes for each mode which can be obtained by performing structural dynamic assessment can be used to localize and classify the severity of the damage [12]. This assessment can be done at various time stages to detect changes in vibrational characteristics to assess the presence of structural damage in the system.

The classical approach in performing SHM is to place an array of sensors and an input force to stimulate the structure. To detect structural changes occurring over time due to damage, vibrational characteristics results must be obtained. The obtained values are compared with previously obtained structural analysis data (undamaged structure). Using this classical approach, variations in structural characteristics can be determined, and the assessment process presented in Figure 10.2 can be fully implemented. However, as previously stated, this approach is labor intensive and costly to implement.

In the literature, there is a preference to use direct data such as using frequency response functions (FRFs), because frequency measurements are often reliable and can be conducted quickly at low costs [3, 11, 23]. For instance, a structure under assessment will demonstrate a drop in eigenfrequencies, indicating a drop in structural stiffness of the system. This suggests the presence of damage in the structure. In the literature, drops in eigenfrequencies have been attributed to failure of the structural support, overload of applied forces causing failure and crack growth in the structure [24–26]. Conversely, increase in FRF frequency peaks (i.e. eigenfrequencies) would indicate a higher than normal stiffness in the structure. To determine the occurrence of damage in the structure with a high level of certainty, it is necessary to detect a 5% natural frequency shift [12]. Other vibrational characteristics that would be indicative of the damage in the structure are the modal shapes and damping values. However, one of the disadvantages of using an FRF is its inability to measure the applied forces on the structure during operation. Therefore, other methods that are not reliant on measurements of the applied forces can be greatly beneficial in performing SHM. This chapter will utilize transmissibility functions to perform a comprehensive SHM. Transmissibility functions are beneficial for use in SHM as they are defined as the quotient of two determined responses in the structure. Therefore, the transmissibility function becomes independent of the applied loads on the system. In previous research, use of the transmissibility functions to perform SHM has demonstrated results with a high degree of confidence [9, 27].

In aerospace applications, damage and fault detections on aeroengines is a routinely scheduled inspection, which involves a time consuming and labor-intensive disassembly process that can be significantly costly [10]. Therefore, another approach and methodology could be beneficial in performing critical safety inspections. Various techniques exist to perform experimental nondestructive testing such as multi-point visual inspection, inspection using ultrasound, and use of X-rays and gamma rays which has to be only carried by experienced personnel [3, 11]. However, these methods have proven to be labor intensive and significantly costly.

In this chapter, bond graph methodology will be implemented to perform a comprehensive theoretical SHM analysis on an aircraft aeroengine. Bond graph methodology can be implemented on various types of engineering systems, ranging from electrical to mechanical systems [28, 29]. Bond graph theory is based on the fact that an engineering system relies on the flow of information and energy. Bond graph theory has been extensively used by researchers and implemented in various engineering applications, including vibration transfer path analysis to study vibration energy propagation in engineering systems [30, 31]. Using bond graph theory, the components of an engineering system can be presented in graphical format. Bond graph modeling is a common way of presenting the essential characteristics of an engineering structure [29]. By implementing bond graph theory, the dynamic equations of motion of the engineering system are readily obtained. Bond graph methodology benefits from advanced and sophisticated theoretical capabilities, making it an extremely useful technique in analyzing structural design modifications prior to manufacturing to reduce development costs [31]. Based on design requirements determined by the integration of the SHM concept with bond graph theory, design guidelines to minimize damage formation and propagations in the engineering structure can be quickly determined at the lowest possible cost. As previously mentioned, the scheduled aeroengine inspections for damage and defect detections are costly to perform. Therefore, it is of great benefit and importance to quickly localize the defect in the aeroengine. It should be mentioned that in the aerospace industry, one of the major requirements and important targets is to implement strategies and protocols that lead to overall reduction and minimization of time and overall costs for development of aircraft components, such as the aeroengine [32]. Moreover, the obtained knowledge from SHM in this chapter will be used to improve on the early design of the aeroengine structure to anticipate possible defects and to propose possible design guidelines to minimize the formation and creation of defects in the structure during the operational lifetime of the aeroengine at the lowest possible cost.

10.2 Fundamentals of Transmissibility Functions

The SHM analysis presented in this chapter will be based on the concept of transmissibility functions to perform detection and localization of damage for the proposed engineering system. Most common approaches use the method in which the dynamic properties are determined based on input and output of the system. For a damped system (k degree of freedom [DOF]), the forced vibration governing dynamic equations of motion is defined by:

$$[M]\ddot{x}(t) + [C]\dot{x}(t) + [K]x(t) = f(t) \tag{10.1}$$

where $x(t)$ is the displacement, $\dot{x}(t)$ is the velocity and $\ddot{x}(t)$ is the acceleration in the system. The applied external force vector on the system is defined as $f(t)$. $[M]$, $[C]$, and $[K]$ are the mass, damping and stiffness parameters of the structure, respectively. In this system, external applied forced is assumed to be applied at the mth DOF; therefore, the force vector has only a single non-zero input. This can be represented by $f(t) = \{0_1, 0_2, 0_3, 0_4, \dots, f_m(t), \dots, 0_k\}^T$, where k is the DOF in the system. Using fast Fourier transform (FFT), Eq. (10.1) can be reformulated to the frequency domain, which is represented by the following formulation:

$$X(\omega) = H(\omega)F(\omega) \tag{10.2}$$

where $H(\omega)$ is defined as the FRF of the system. Figure 10.3 illustrates a single input and single output (SISO). Again, assuming the force vector has only a single non-zero input at the mth DOF, the external force vector represented in the frequency domain by $F(\omega)$ can be formulated as:

$$F(\omega) = \{0_1, 0_2, 0_3, 0_4, \dots, F_m(\omega), \dots, 0_k\}^T \tag{10.3}$$

Transmissibility function in the system of interest is determined by determining the quotient of two frequency response spectra in the system. The frequency response spectra in this case are defined as the acceleration vector in the frequency domain, which can be formulated as:

$$A(\omega) = -\omega^2 H(\omega)F(\omega) \tag{10.4}$$

The transmissibility function between the output DOF i and j can be defined by T_{ij}, which can be determined by finding the ratio between $A_i(\omega)$ and $A_j(\omega)$. Similar to FRF calculations, the transmissibility function can be determined using a nonparametric approach. The transmissibility function can be formulated as:

$$T_{ij}(\omega) = \frac{A_{im}(\omega)}{A_{jm}(\omega)} = \frac{X_{im}(\omega)}{X_{jm}(\omega)} = \frac{-\omega^2 H_{im}(\omega)F(\omega)}{-\omega^2 H_{jm}(\omega)F(\omega)} = \frac{H_{im}(\omega)}{H_{jm}(\omega)} \tag{10.5}$$

Figure 10.3 Engineering system dynamic vibration illustration for SISO.

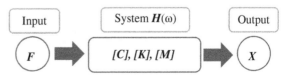

where H_{im} and H_{jm} are the elements of the FRF matrix. X_{im} and X_{jm} are the displacements for the DOF of interest. It can be observed from Eq. (10.5) that the transmissibility function is inherently defined as a dynamic characteristic of the system, defined by the ratio of the FRFs (receptance matrix) in the system for the points of interest. However, as evident from Eq. (10.5), it can be noted that the transmissibility function is defined based on the quotient of the frequency response spectra without the need of considering the input forces or the system FRFs. The frequency response spectra in the system can be defined by acceleration ($A(\omega)$) or displacement ($X(\omega)$) of the system defined for the DOF of interest.

10.3 Bond Graphs

As previously discussed, bond graph methodology can be implemented to represent a wide range of engineering systems. Bond graph methodology has been implemented by researchers for various engineering and research applications, including system modeling and development, system dynamic analysis, vibration analysis, and transfer path analysis. The systematic nature and theoretical capabilities of bond graph methodology allows the implementation of the concept for various research applications.

10.3.1 Bond Graph Theory

To theoretically determine the required transmissibility functions to perform SHM, the concept and theory of creating bond graphs must be developed and elaborated on. In bond graph methodology, the engineering structure under analysis is broken down into smaller parts that are termed "components" in the system. For graphical representation of an engineering system, subsystems of an engineering system are connected through ports, in which power can flow between them. Power variables are classified as either flow ($f(t)$) or effort ($e(t)$) variables [29]. Effort would encompass variables such as force, torque, pressure, and electrical voltage depending on the nature and type of the engineering system under study. By implementing the bond graph approach, a hybrid model can be developed for which all types of power variables from all engineering disciplines are incorporated. In bond graph methodology, the flow variable would include variables such as velocity, angular velocity, volumetric flow rate, and electrical current.

In bond graph methodology, the power ($P(t)$) is defined as the multiplication of the effort and flow variables, defined by:

$$P(t) = e(t).f(t) \tag{10.6}$$

The energy variable is defined by the integration of the effort or flow variable present in the system, which can be formulated as:

$$E(t) = \int^t e(t)dq(t) = \int^t f(t)dp(t) \tag{10.7}$$

Please refer to Chapter 9 for a full and comprehensive explanation regarding bond graph theory.

Figure 10.4 Power bond representation to link the subsystem components.

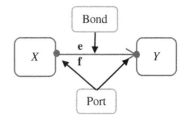

10.3.2 Graphical Representation and Modeling of Bond Graphs

To perform SHM analysis using bond graph methodology, it is first necessary to develop the bond graph representation of the selected engineering structure. A bond graph model is composed of components connected through lines in which energy can transfer through them. Each bond can represent both power and effort variables, as illustrated in Figure 10.4.

There are two forms of junctions which are utilized in the development of bond graphs. The first type is the 0-junction and the second is the 1-junction. Examples of a 0-junction and 1-junction are shown for a 2-bond element in Figures 10.5 and 10.6, respectively.

$$——— \; 0 \; ———$$

Figure 10.5 Graphical model of a 2-bond 0-junction element.

At both the 0- and 1-junctions, the conservation of energy principle is applicable. This is shown by the following equation:

$$——— \; 1 \; ———$$

Figure 10.6 Graphical model of a 2-bond (10.8) 1-junction element.

$$e_1 f_1 + e_2 f_2 + e_3 f_3 + \ldots + e_n f_n = 0$$

The following principle is applicable to the 0-junction node. By mathematical manipulation, the following equation would indicate that all the effort variables are the same at the 0-junction element.

$$f_1(t) + f_2(t) + f_3(t) + \ldots + f_n(t) = 0 \tag{10.9}$$

The following principle is applicable to the 1-junction node.

$$e_1(t) + e_2(t) + e_3(t) + \ldots + e_n(t) = 0 \tag{10.10}$$

For a full and comprehensive explanation of the bond graph modeling concept, please refer to Chapter 9.

10.3.3 Determination of State-Space Equations Using Bond Graph Theory

After developing bond graph representation of the engineering structure, the state-space equation of the system based on significant variables can be determined [29]:

1. The significant input and energy variables in the system must be determined.
2. Utilizing the selected significant variables, a first-order set of equations can be determined.
3. The obtained equations are formulated in terms of state-space forms

Using the outlined approach, the following set of equations can be determined:

$$\dot{x}_1(t) = a_{11}x_1 + a_{12}x_2 + \ldots + a_{1n}x_n + b_{11}u_1 + b_{12}u_2 + \ldots + b_{1r}u_r$$
$$\dot{x}_2(t) = a_{21}x_1 + a_{22}x_2 + \ldots + a_{2n}x_n + b_{21}u_1 + b_{22}u_2 + \ldots + b_{2r}u_r$$
$$\dot{x}_3(t) = a_{31}x_1 + a_{32}x_2 + \ldots + a_{3n}x_n + b_{31}u_1 + b_{32}u_2 + \ldots + b_{3r}u_r \tag{10.11}$$
$$\vdots$$
$$\dot{x}_n(t) = a_{n1}x_1 + a_{n2}x_2 + \ldots + a_{nn}x_n + b_{n1}u_1 + b_{n2}u_2 + \ldots + b_{nr}u_r$$

where a_{ij} and b_{ij} are constants.

Using a matrix formation, Eq. (10.11) can be formulated as (in the time domain):

$$\dot{X} = AX + BU \tag{10.12}$$

$$Y = CX + DU \tag{10.13}$$

where X is the vector of state variables, U is the input vector and Y is the output vector. Matrix A and B are the state matrix and the input matrix, respectively. Matrix C defines the output matrix and matrix D defines the direct transmission matrix. Equations (10.12, 10.13) can be transformed to the frequency domain analysis using the Laplace transform:

$$X = (sI - A)^{-1}BU \tag{10.14}$$

$$Y = [C(sI - A)^{-1}B + D]U \tag{10.15}$$

For a comprehensive explanation to determine the state-space equations using bond graph theory, please see Chapter 9.

10.3.4 Determination of Transmissibility Functions Using the Bond Graph Concept

As discussed in the previous section, by obtaining the dynamic equations of motion, the transmissibility function between various DOFs in the structure (aeroengine) can be determined for various inertia elements. This is achieved by implementing the theory of global transmissibility. The global transmissibility between DOF i and j is formulated as the ratio of displacement between these DOFs when an applied load is exerted on i and there are no other restrictions on the system [28]. Using Laplace transform, the transmissibility function presented in Eq. (10.5) can be reformulated as [28]:

$$T_{ij}(\omega) = \frac{SX_j}{SX_i} = \frac{f^{(j)}}{f^{(i)}} \tag{10.16}$$

where $f^{(i)}$ demonstrates the flow in the respective element. It should be noted that Eq. (10.16) is dependent on the mth DOF where the external load is imposed on the structure. $f^{(i)}$, which represents the flow to the element, is formulated as [28]:

$$f^{(i)} = [C^i(sI - A)^{-1}B + D^i]U \tag{10.17}$$

This leads to the following equation:

$$T_{ij}(\omega) = \frac{[C^j(sI - A)^{-1}B + D^j]U}{[C^i(sI - A)^{-1}B + D^i]U} \tag{10.18}$$

Using the Bode magnitude function which shows the frequency response of the transmissibility at various frequencies [28], the global transmissibility function can be formulated as:

$$T_{ij}(\omega) = \frac{\text{Bodemag}(A, b, C^j, D^j)}{\text{Bodemag}(A, b, C^i, D^i)} \tag{10.19}$$

10.4 Structural Health Monitoring Damage Indicator Factors

The transmissibility function for a healthy structure between the DOF of interest can be represented by Eq. (10.16). The same way that a healthy structure is modeled using Eq. (10.16), the damaged structure can be modeled using the same approach:

$$T_{ij}(\omega)^{DMG} = \frac{SX_j^{DMG}}{SX_i^{DMG}} = \frac{f^{(j)DMG}}{f^{(i)DMG}} = \frac{\text{Bodemag}(A, b, C^j, D^j)^{DMG}}{\text{Bodemag}(A, b, C^i, D^i)^{DMG}} \tag{10.20}$$

Due to the presence of structural damage and defects, variations in the dynamic properties of the engineering system can be expected [12, 24–26]. Therefore, it can be anticipated that variation in the dynamic property and characteristics of the structure are reflected in the determined transmissibility transfer functions of the damaged and defective structure.

To localize the damage in the engineering structure, the transmissibility transfer functions of close and adjacent elements (DOF) will be determined for both the damaged and undamaged structure. The obtained transmissibility function with the greatest degree of change regarding the obtained transmissibility transfer function when comparing the damaged and the defective structure will pinpoint and localize the damage in the structure.

To achieve damage detection and localization, it is first important to utilize, implement and propose a damage indicator (DI) concept. Various techniques have been investigated in the literature [2, 3, 33–35]. One of the DI concepts proposed by Sampaio et al. [27] proposes the calculation of peak amplitude differences between the damaged and the healthy structure as the DI in the structure. This can be formulated by the following equation [23]:

$$D_{ij}^{IND}(\omega) = \sum_{\omega} |T_{ij}(\omega) - T_{ij}(\omega)^{DMG}| \tag{10.21}$$

By using this approach, $D_{ij}^{IND}(\omega)$ computes a single value for each transmissibility function for the location of interest in the structure. The transmissibility functions are computed for all neighboring locations on the structure. Using Eq. (10.21), the DI would lead to the highest $D_{ij}^{IND}(\omega)$ value for the location with defect [23]. A similar indicator has been developed by Johnson et al. [9].

$$D_{ij}^{IND}(\omega) = \left| \sum_{\omega} \left(1 - \frac{T_{ij}(\omega)^{DMG}}{T_{ij}(\omega)} \right) \right| \tag{10.22}$$

10.5 Aircraft Aeroengine Parametric Modeling

As previously discussed, the proposed SHM methodology that is implemented in conjunction with bond graph methodology will be implemented on a proposed reduced and

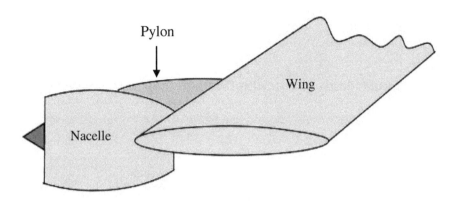

Figure 10.7 Aeroengine wing assembly.

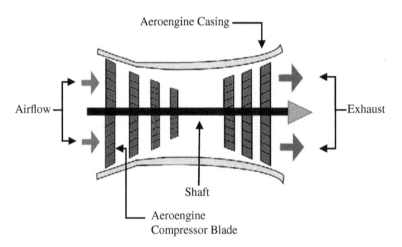

Figure 10.8 Simplified aeroengine cross-section.

simplified aeroengine model. The reduced aeroengine model is illustrated in Figure 10.7. The illustrated aeroengine model consists of a rotating shaft that is attached to compressor blades that work in various compression stages. The shaft and the blades are housed in several casing structures that are attached together by flange connections. Due to the high operating conditions of the aeroengine, the casings are manufactured from high temperature resistant alloys [36], such as titanium. The aeroengine casings are housed in the nacelle which itself is attached to the pylon. The pylon is attached to the aircraft wing. Figure 10.8 represents the cross-section of a typical aeroengine.

To perform SHM analysis on an aircraft aeroengine, a reduced aeroengine lumped model is proposed in Figure 10.9. The proposed aeroengine model consists of a rotor system. The rotor system imposes external excitation loadings on the aeroengine components due to rotor dynamic unbalance forces initiating from the rotor assembly. These unbalance forces are transferred from the rotor system assembly to the aeroengine casings. These external forces are transferred to the casings through two parallel spring and damper systems (C_{AL}, K_{AL}, C_{AR}, and K_{AR}). The aeroengine casing is defined by M_{Casing} (mass), M_I (inertia), and M_M (modal mass). The aeroengine casing is attached to the nacelle through a single pair damper and spring system (C_B and K_B) to represent the damping and stiffness of values of the aeroengine connection to the nacelle, which is represented by $M_{Nacelle}$. The nacelle is attached to the wing (represented by ground due to its high relative mass when it has been fueled) through the aeroengine pylon. The pylon is represented by its structural stiffness, K_P.

The bond graph of the proposed reduced aeroengine model is illustrated in Figure 10.10. The bond graph of the aeroengine is developed based on the theory presented in Section 10.3.2. The physical characteristics of the reduced aeroengine model are presented in Table 10.1. Considering the developed bond graph model, the governing dynamic equations of motion are determined based on the methodology presented in Section 10.3.3. The state variable vector of the aeroengine model (x) is defined by Eq. (10.23).

$$
\begin{bmatrix} \dot{q}3 \\ \dot{q}12 \\ \dot{q}22 \\ \dot{P}23 \\ \dot{q}27 \\ \dot{P}35 \\ \dot{P}36 \\ \dot{P}38 \end{bmatrix} =
\begin{bmatrix}
\frac{-K_{AL}}{C_{AL}} & 0 & 0 & 0 & 0 & 0 & 0 & 0 \\
0 & \frac{-K_{AR}}{C_{AR}} & 0 & 0 & 0 & 0 & 0 & 0 \\
0 & 0 & 0 & \frac{1}{M_{Nacelle}} & 0 & 0 & 0 & 0 \\
0 & 0 & -K_P & \frac{-C_B}{M_{Nacelle}} & -K_B & \frac{-C_B}{Z0(XA)M_{Casing}} & \frac{-C_B}{Z2(XA)J} & \frac{-C_B}{Y1(da)M_{Modal}} \\
0 & 0 & 0 & \frac{-1}{M_{Nacelle}} & 0 & \frac{-1}{Z0(XA)M_{Casing}} & \frac{-1}{Z2(ZA)J} & \frac{-1}{Z1(XA)M_{Modal}} \\
0 & 0 & 0 & \frac{-C_B}{Z2(XA)M_{Nacelle}} & \frac{K_B}{Z3(XA)} & \frac{-C_B}{Z0(XA)M_{Casing}} & \frac{-C_B}{Z2(XA)J} & \frac{-C_B}{Z1(XA)M_{Modal}} \\
0 & 0 & 0 & \frac{-C_B}{Z0(XA)M_{Nacelle}} & \frac{K_B}{Z0(XA)} & \frac{-C_B}{Z0(XA)M_{Casing}} & \frac{-C_B}{Z2(XA)J} & \frac{-C_B}{Z1(XA)M_{Modal}} \\
0 & 0 & 0 & \frac{-C_B}{Z1(XA)M_{Nacelle}} & \frac{K_B}{Z1(XA)} & \frac{-C_B}{Z0(XA)M_{Casing}} & \frac{-C_B}{Z2(XA)J} & \frac{-C_B}{Z1(XA)M_{Modal}}
\end{bmatrix}
\begin{bmatrix} q3 \\ q12 \\ q22 \\ P23 \\ q27 \\ P35 \\ P36 \\ P38 \end{bmatrix}
$$

$$
+ \left[\frac{Se_{Rotor\,L}}{CAL} \quad \frac{Se_{Rotor\,R}}{CAL} \quad 0 \quad 0 \quad 0 \quad \frac{Se_{Rotor\,L}}{Z0(XR)} + \frac{Se_{Rotor\,L}}{Z0(XL)} \quad \frac{Se_{Rotor\,R}}{Z2(XR)} + \frac{Se_{Rotor\,L}}{Z2(XL)} \quad \frac{Se_{Rotor\,R}}{Z1(XR)} + \frac{Se_{Rotor\,L}}{Z1(XL)} \right]^T
$$

$$(10.23)$$

10.6 Results and Discussion

To perform SHM on the proposed aeroengine model, the global transmissibility transfer function is determined for various points of interest in the aeroengine. To implement the proposed SHM concept, the proposed aeroengine lumped model is categorized into three regions, which are illustrated in Figure 10.11. This is done to allow the follow through with

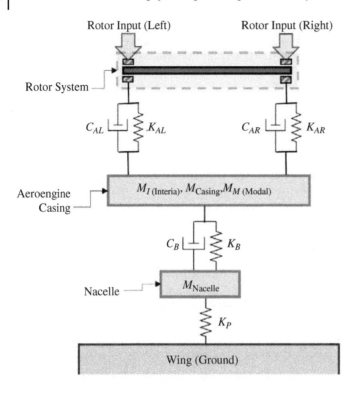

Figure 10.9 Reduced aeroengine lumped model representation.

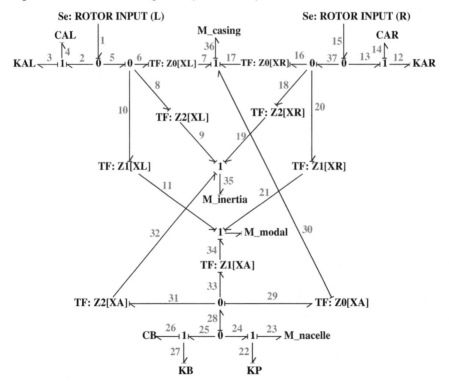

Figure 10.10 Reduced aeroengine bond graph model.

Table 10.1 Reduced aeroengine model parameters.

Parameter	Assigned value
M_b	315 kg
Casing density (ρ)	4500 kg/m³
Casing modal mass (M_1, modal)	255.3 kg
Casing mass moment of inertia (M_1, inertia)	26.4 kg/m²
Casing length	1 m
Casing width	1 m
Casing height	0.07 m
M_{Nacelle}	500 kg
K_{AL}	6×10^4 N/m
K_{AR}	6×10^4 N/m
K_B	4×10^5 N/m
K_P	7×10^6 N/m
C_{AL}	5 N.s/m
C_{AR}	5 N.s/m
C_B	4 N.s/m

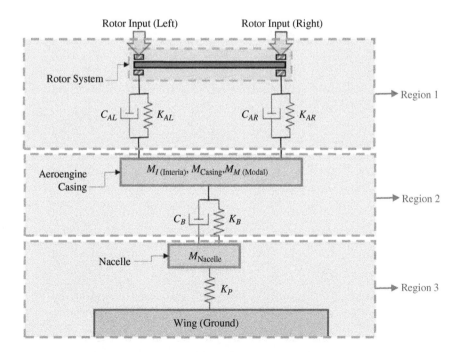

Figure 10.11 Categorized reduced aeroengine lumped model.

SHM assessment phases illustrated in Figure 10.2. The aeroengine categorization is done to improve structural referencing and localization of the damage in this chapter.

Using Eq. (10.18), the transmissibility transfer function can be calculated for various DOFs and points of interest on the proposed aeroengine model for frequency range of ω_1–ω_2. In this section, the global transmissibility between the rotor input (bond 13) and the aeroengine wing (bond 23) is determined. For comprehensive and detailed explanation of procedure in obtaining the output matrix ($\mathbf{C}$) and direct transmission matrix ($\mathbf{D}$), please refer to Chapter 9. The global transmissibility between bonds 23 and 13 is defined by $\frac{f23}{f13}$.

For bond 13, using Eqs. (10.12) and (10.13) and Eq. (10.23), the output and direct transmission matrices can be defined as:

$$Y_{13} = \mathbf{C_{13}}X + \mathbf{D_{13}}\vec{U} = \frac{Se\,(\text{Input Right}) - q15.KAR}{CAR} \tag{10.24}$$

For bond 23, the output and direct transmission matrices are defined as:

$$Y_{23} = \mathbf{C_{23}}X + \mathbf{D_{23}}\vec{U} = \frac{P23}{M_{\text{Nacelle}}} \tag{10.25}$$

Using Eqs. (10.24) and (10.25), the global transmissibility transfer function is defined as $T_{23,13}(\omega) = \frac{\text{Bodemag}(A,B,C^{23},D^{23})}{\text{Bodemag}(A,B,C^{13},D^{13})}$ based on Eq. (10.19). The global transmissibility transfer function assigned between the aeroengine rotor and the wing is shown in Figure 10.12.

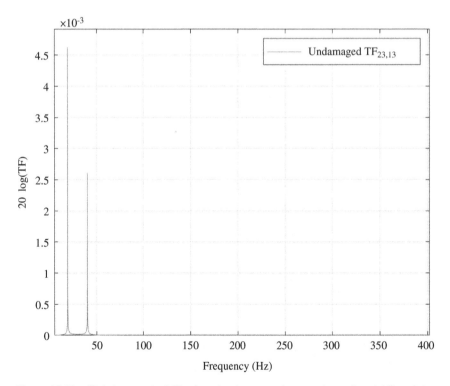

Figure 10.12 Global transmissibility function between the rotor input (bond 13) and the aeroengine wing (bond 23).

Figure 10.12 illustrates the FRF of the undamaged aeroengine structure. The parameters implemented in this analysis are based on the parameters provided in Table 10.1. SHM and damage detection is performed using the obtained transmissibility transfer functions determined by the bond graph analysis.

Initially, to confirm the occurrence of damage within the structure, the global transmissibility transfer function (previously determined, bond 13 and bond 23) is determined for a damaged aeroengine model. The introduced damage intends to represent loss of structural stiffness that is caused by crack initiation and growth at the connection points [37] due to fatigue. To represent a propagation of crack and damage in the region of interest in the aeroengine, 8, 10, and 12% stiffness reduction is introduced in the structure for the identified region within the structure [1]. The global transmissibility transfer function is determined for both the damaged and undamaged structure. Any change in FRF frequency peak amplitudes and changes in eigenfrequencies is indicative of damage in the structure [24–26]. Table 10.2 illustrates the implemented parameters that represent the structural damage at the attachments points between the rotor system and the casings (K_{AL} and K_{AR}). The obtained FRF for a 10% reduction in K_{AL} and K_{AR} is illustrated in Figure 10.13.

As it can be observed from Figure 10.13, a 10% reduction in K_{AL} and K_{AR} results in a drop in the global transmissibility transfer function peak amplitude. The second FRF peak (40 Hz) illustrates the highest amount of amplitude change, which is calculated to be approximately 20%. This is indicative of damage and defect in the structure and the attachment points connecting the rotor and the casing. The connection between the rotor and the aeroengine casing would include the squeezed filmed dampers which are used to attenuate the aeroengine rotor vibrations. From this analysis, it can be concluded that any significant detection of changes solely for transmissibility peak would be indicative of damage at the attachment points between the rotor system and the casings. In addition, the obtained results can be used by design engineers to theoretically predict dynamic behavior of a damaged aeroengine structure and anticipate possible dynamic structural behavior due to damage and propose design guidelines to avoid catastrophes. For instance, if a shift in eigenfrequencies is anticipated for small defects and damage in the structure, design guidelines can be proposed that would eliminate the possibility that the shifted eigenfrequency of the structure might coincide with the operational frequency of the aeroengine, which could lead to amplified vibration and noise transfer from the aeroengine. To perform damage localization in the aeroengine structure, the DI factor identified in Section 10.4 is implemented. By determining the transmissibility functions for each of the identified regions in Figure 10.11, the DI value for each region is calculated based on Eq. (10.22). This is done for all the three identified regions of the proposed aeroengine model. In general,

Table 10.2 Modified structural parameters for damaged and undamaged aeroengine structural attachment points (region 1).

Parameter	Undamaged structure	Damaged structure (8% reduction)	Damaged structure (10% reduction)	Damaged structure (12% reduction)
$K_{AL=} K_{AR}$	6×10^4 N/m	5.52×10^4 N/m	5.4×10^4 N/m	5.28×10^4 N/m

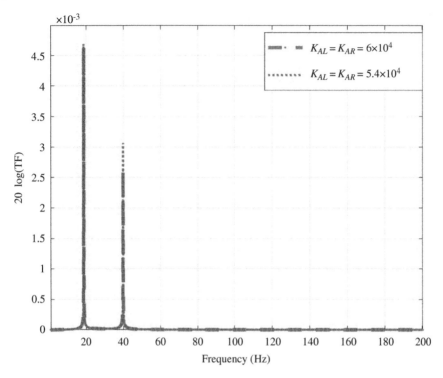

Figure 10.13 Global transmissibility function between the rotor input excitation and the aeroengine wing, for the structural connection K_{AL} and K_{AR} (10% stiffness reduction).

high DI numbers are indicative of damage within that region. Figure 10.14 shows the DI numbers for 8, 10, and 12% damage in the structure (K_{AL} and K_{AR}). All the DI numbers presented in Figure 10.14 are normalized. As it can be observed from Figure 10.14, DI numbers are indicating a high degree of difference in the FRF values, implying significant structural damage in region 1. This analysis indicates that the structural damage is mainly localized to region 1. In addition, it can be observed that region 2 is illustrating higher DI values compared with region 3. This is attributed to region 2 being a close neighbor to region 1, causing the dynamic characteristics of region 2 to be marginally affected [1, 38]. By performing this analysis, the present damage and defect can be readily localized using a fully theoretical approach. By performing the proposed theoretical approach in obtaining DI numbers, the obtained DI values can be compared with experimentally obtained results to analyze the extent and degree of damage in the structure. For instance, the theoretical data can be used to analyze and study the extent of crack propagation in a real structure to analyze and study the remaining lifetime and to assess the safety and reliability of the aeroengine for operation. Therefore, it can be concluded that bond graph methodology utilized in conjunction with SHM principles provides an early theoretical detection tool to detect the extent of structural damage in various engineering structures, including aeroengines. In addition, the obtained DI values can be utilized to propose design guidelines that would minimize the extent of damage propagation over time by proposing structural modifications in the aeroengine.

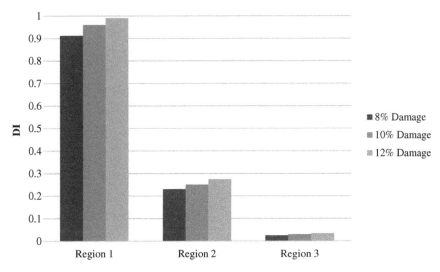

Figure 10.14 DI numbers for various damage characteristics (K_{AL} and K_{AR}).

The proposed SHM analysis is also implemented on the second identified region of the aeroengine. Table 10.3 illustrates the utilized stiffness values representing 8, 10, and 12% reduction of stiffness regarding the attachment point between the aeroengine casing and the nacelle (K_B). The obtained transmissibility transfer function for a 10% reduction in K_B value is illustrated in Figure 10.15. The FRF results are compared with the undamaged structure. As can be observed from Figure 10.15, any defects and damage in region 2 of the aircraft which is reflected in the K_B parameters results in both a shift in eigenfrequencies and change in peak amplitudes. The first transmissibility peak shows a 20% drop in peak amplitude, whereas the second transmissibility peak shows a 7% shift in eigenfrequency and approximately 28% change in transmissibility peak. The illustrated transmissibility function behavior and characteristic can be attributed to damage at region 1, which affects and alters the overall stiffness of the structure, causing a shift in eigenfrequency and drop in transmissibility amplitude [12]. In addition, DI analysis is performed on the aeroengine structure to provide a technique and tool of assessing the extent and degree of damage in physical aeroengine structures by comparing theoretical and experimental results, as previously discussed. DI values are calculated for all three aeroengine regions for 8, 10, and 12 damage at the attachment points between the casing and nacelle (reflected by K_B). As evident from Figure 10.16, DI values indicate a localized damage at region 2. The dynamic

Table 10.3 Modified structural parameters for damaged and undamaged aeroengine structural attachment points (region 2).

Parameter	Undamaged structure	Damaged structure (8% reduction)	Damaged structure (10% reduction)	Damaged structure (12% reduction)
K_B	4×10^5 N/m	3.68×10^5 N/m	3.6×10^5 N/m	3.52×10^5 N/m

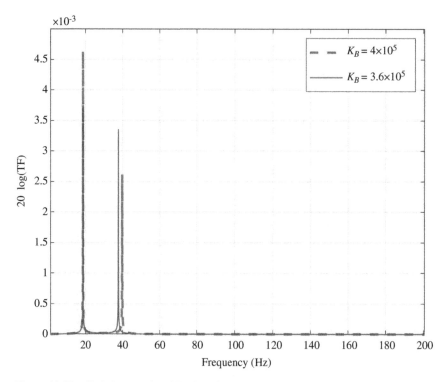

Figure 10.15 Global transmissibility function between the aeroengine casing and the nacelle, for the structural connection K_B (10% stiffness reduction).

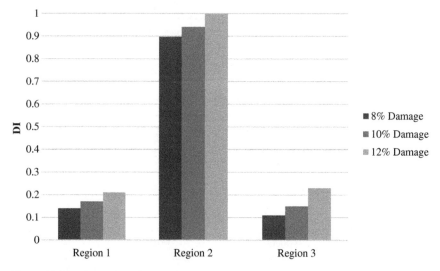

Figure 10.16 DI numbers for various damage characteristics (K_B).

behavior of regions 1 and 3 has also been slightly affected due to change in dynamic characteristics of region 2. The presented DI analysis can be utilized to study the safety and reliability of the aeroengine structure by comparing experimental DI results to approximately predict the extent of damage in the structure.

Similar SHM analysis is implemented on region 3 of the proposed aeroengine model. Table 10.4 illustrates the utilized stiffness parameters, representing stiffness reduction due to damage and defect of the connection between the nacelle and the wing, represented by the aeroengine pylon (K_P). The obtained transmissibility transfer function for a 10% reduction in K_P value (damaged structure) and undamaged structure is illustrated in Figure 10.17. As evident from Figure 10.17, it can be observed that damage at region 3 of the aeroengine results is in both eigenfrequency shift (5%) and drop in amplitude (20%) for

Table 10.4 Modified structural parameters for damaged and undamaged aeroengine structural attachment points (region 3).

Parameter	Undamaged structure	Damaged structure (8% reduction)	Damaged structure (10% reduction)	Damaged structure (12% reduction)
K_P	7×10^6 N/m	6.44×10^6 N/m	6.3×10^6 N/m	6.16×10^6 N/m

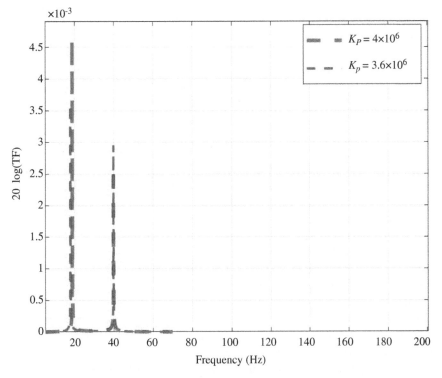

Figure 10.17 Global transmissibility function between the aeroengine casing and the nacelle, for the structural connection K_p (10% stiffness reduction).

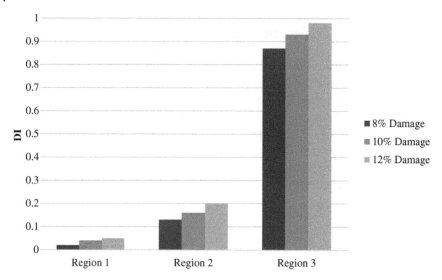

Figure 10.18 DI numbers for various damage characteristics (K_p).

the first peak. The second peak does not illustrate significant change in peak values or shift in eigenfrequencies. The illustrated transmissibility transfer function is representative of damage at the third identified region of the aeroengine. The obtained results can be used to seamlessly identify the type, extent, and characteristic of the damage within an aeroengine, when compared with experimental results. DI values are calculated for all three identified regions for 8, 10, and 12% damage at the attachment points between the nacelle and the aircraft wing (reflected by K_p). The DI results are presented in Figure 10.18.

10.7 Conclusion

This chapter has proposed a comprehensive and novel study of SHM analysis in conjunction with bond graph theory for implementation on large structures, specifically aeroengines. This study has demonstrated the advantage and superiority of using bond graph methodology in conjunction with SHM analysis to propose theoretical guidelines that would reduce operational costs, such as reducing the detection and maintenance costs while providing theoretical tools to assess the safety of the structure by providing an early detection tool to detect structural damage. This study has shown that bond graph theory can be used in conjunction with SHM methodology to propose a transmissibility-based theoretical approach in determining DI factors that would enable the localization of damage within the structure using calculated transmissibility transfer functions. The determined DI values illustrated high reliability and efficiency in localizing the damage within the aeroengine structure. The obtained DI values also provide a theoretical basis to study and assess the extent of damage in an actual aeroengine prototype. It has also been shown that proposed methodology can be used to detect the presence of damage quickly, when compared with other tools proposed in the literature. In addition to the improved safety and performance, the obtained knowledge from the proposed SHM analysis can be utilized to implement early design modification and guidelines based on the predetermined safety factors of the aeroengine, significantly improving reliability and operational life time of the aeroengine.

References

1 Feng, L., Yi, X., Zhu, D. et al. (2015). Damage detection of metro tunnel structure through transmissibility function and cross correlation analysis using local excitation and measurement. *Mech. Syst. Sig. Process* 60: 59–74.

2 Chesné, S. and Deraemaeker, A. (2013). Damage localization using transmissibility functions: a critical review. *Mech. Syst. Sig. Process* 38: 569–584.

3 Maia, N.M.M., Almeida, R.A.B., Urgueira, A.P.V., and Sampaio, R.P.C. (2011). Damage detection and quantification using transmissibility. *Mech. Syst. Sig. Process* 25: 2475–2483.

4 Schallhorn, C. and Rahmatalla, S. (2015). Crack detection and health monitoring of highway steel-girder bridges. *Struct. Heal Monit.* 14: 281–299.

5 Caccese, V., Mewer, R., and Vel, S.S. (2004). Detection of bolt load loss in hybrid composite/metal bolted connections. *Eng. Struct.* 26: 895–906.

6 Hemez, F.M. and Doebling, S.W. (2001). Review and assessment of model updating for non-linear, transient dynamics. *Mech. Syst. Sig. Process* 15: 45–74.

7 Ou, J. and Li, H. (2009). Structural health monitoring in the Mainland of China: review and future trends. *Struct. Heal Monit.* From Syst Integr to Auton Syst - Proc 7th Int Work Struct Heal Monit IWSHM 2009 1: 29–41.

8 Güemes, A. (2013). SHM technologies and applications in aircraft structures. 5th International Symposium on NDT in Aerospace, Singapore.

9 Johnson, T.J. and Adams, D.E. (2002). Transmissibility as a differential indicator of structural damage. *J. Vib. Acoust. Trans. ASME* 124: 634–641.

10 Fritzen, C.P. (2010). Vibration-based techniques for structural health monitoring. *Struct. Heal Monit.* https://doi.org/10.1002/9780470612071.ch2.

11 Sampaio, R.P.C. and Maia, N.M.M. (2009). Strategies for an efficient indicator of structural damage. *Mech. Syst. Sig. Process* 23: 1855–1869.

12 Salawu, O.S. (1997). Detection of structural damage through changes in frequency: a review. *Eng. Struct.* 19: 718–723.

13 Sabzi, G.A., Davoodi, M., and Mohamadzade, R.M. (2019). Structural damage detection using frequency response function. *Int. J. Eng. Technol.* 11: 37–59.

14 Ribeiro, A.M.R., Silva, J.M.M., and Maia, N.M.M. (2000). On the generalisation of the transmissibility concept. *Mech. Syst. Sig. Process* 14: 29–35.

15 Doebling, S.W.S., Farrar, C.R.C., Prime, M.B.M., and Shevitz, D.W.D. (1996). Damage identification and health monitoring of structural and mechanical systems from changes in their vibration characteristics: a literature review. *Los Alamos Natl. Lab.* https://doi .org/10.2172/249299.

16 Farrar, C., Farrar, C., Hemez, F. et al. (2003). Damage Prognosis: Current Status and Future Needs. *Tech. Rep.*

17 Chelidze, D. (2003). Dynamical systems approach to material damage diagnosis. In: *Proc ASME Des Eng Tech Conf.* ASME. https://doi.org/10.1115/DETC2003/VIB-48452.

18 Wedman, S. and Wallaschek, J. (2001). The application of a lifetime observer in vehicle technology. *Key Eng. Mater.* 204–205: 153–162.

19 Wallaschek, J., Wedman, S., and Wickord, W. (2002). Lifetime observer: an application of mechatronics in vehicle technology. *Int. J. Veh. Des.* 28: 121–130.

20 Natke, H. and Cempel, C. (1997). *Model-Aided Diagnosis of Mechanical Systems*. Springer. https://doi.org/10.1007/978-3-642-60413-3.

21 Fritzen, C.-P., Jennewein, D., and Kiefer, T. (1998). Damage detection based on model updating methods. *Mech. Syst. Signal Process* 12: 163–186.

22 Link, M. (1999). Updating of analytical models – basic procedures and extensions. In: *Proceedings of the NATO ASI*, 281–304. Kluwer.

23 Devriendt, C., Presezniak, F., De Sitter, G. et al. (2010). Structural health monitoring in changing operational conditions using tranmissibility measurements. *Shock Vib.* 17: 651–675.

24 Zhu, H.P., He, B., and Chen, X.Q. (2005). Detection of structural damage through changes in frequency. *Wuhan Univ. J. Nat. Sci.* 10: 1069–1073.

25 Dewolf, J.T. (1990). Experimental study of bridge monitoring technique. *J. Struct. Eng.* 116: 2532–2549.

26 Salane, B.H.J. (1990). Identification of modal properties of bridges. *J. Struct. Eng.* 116: 2008–2021.

27 Sampaio, R.C., Silva, J.M.M., and Technical, A.M.R.R. (2001). Transmissibility techniques for damage detection. In: *19th International Modal Analysis Conference*. https://doi.org/10.13140/2.1.3720.8007.

28 Mashayekhi, M.J. and Behdinan, K. (2015). Analytical transmissibility based transfer path analysis for multi-energy-domain systems using bond graphs. *J. Vib. Control* https://doi.org/10.1177/1077546317696362.

29 Karnopp, D.C., Margolis, D.L., and Rosenberg, R.C. (2006). *System Dynamics*, 4e. Wiley.

30 Mir-Haidari, S.-E. and Behdinan, K. (2019). Aero-engine vibration propagation analysis using bond graph transfer path analysis and transmissibility theory. In: *Proceedings of the ASME 2019 International Mechanical Engineering Congress and Exposition*. ASME. https://doi.org/10.1115/IMECE2019-10773.

31 Mir-Haidari, S.E. and Behdinan, K. (2019). On the vibration transfer path analysis of aero-engines using bond graph theory. *Aerosp. Sci. Technol.* 95: 105516.

32 Yin, F., Gangoli Rao, A., Bhat, A., and Chen, M. (2018). Performance assessment of a multi-fuel hybrid engine for future aircraft. *Aerosp. Sci. Technol.* 77: 217–227.

33 Maia, N.M.M., Ribeiro, A.M.R., Fontul, M. et al. (2007). Using the detection and relative damage quantification indicator (DRQ) with transmissibility. *Key Eng. Mater.* 347: 455–460.

34 Serafini, J., Bernardini, G., Porcelli, R., and Masarati, P. (2019). In-flight health monitoring of helicopter blades via differential analysis. *Aerosp. Sci. Technol.* 88: 436–443.

35 Diamanti, K. and Soutis, C. (2010). Structural health monitoring techniques for aircraft composite structures. *Prog. Aerosp. Sci.* 46: 342–352.

36 Bettebghor, D., Blondeau, C., Toal, D., and Eres, H. (2013). Bi-objective optimization of pylon-engine-nacelle assembly: weight vs. tip clearance criterion. *Struct. Multidiscip. Optim.* 48: 637–652.

37 Friswell, M.I. and Penny, J.E.T. (2002). Crack modeling for structural health monitoring. *Struct. Heal Monit.* 1: 139–148.

38 Yan, W.J., Zhao, M.Y., Sun, Q., and Ren, W.X. (2019). Transmissibility-based system identification for structural health monitoring: fundamentals, approaches, and applications. *Mech. Syst. Sig. Process.* 117: 453–482.

Index

Advanced Multifunctional Lightweight Aerostructures: Design, Development, and Implementation,
First Edition. Kamran Behdinan and Rasool Moradi-Dastjerdi.
© 2021 John Wiley & Sons Ltd. This Work is a co-publication between John Wiley & Sons Ltd and ASME Press.